AF760344

MECHANICAL ENGINEERING

MECHANICAL ENGINEERING

Abe Montgomery Addison

Apple Academics

MECHANICAL ENGINEERING

Copyright © Apple Academic Press Inc.

All rights reserved. No part of this publication may be reproduced or transmitted in any form or by any means, electronic or mechanical, including photocopying, recording, taping, or any information storage and retrieval system, without permission from the publisher.

Published by
Apple Academic Press Inc.
3333 Mistwell Crescent
Oakville, ON L6L 0A2
Fax: (866) 222-9549
www.appleacademicpress.com

ISBN 978-0-9811660-5-6

Abe Montgomery Addison

First Edition: 2010

9 8 7 6 5 4 3 2 1

Library and Archives Canada Cataloguing in Publication Data
CIP Data on file with the Library and Archives Canada

Printed in India.

To the best of the publisher's knowledge, all articles in this book are copyrighted to the individual authors and are licensed and distributed under the Creative Commons Attribution License, which permits unrestricted use, distribution, and reproduction in any medium, provided the original work is properly cited.

CONTENTS

INTRODUCTION

Mechanical engineering is an engineering discipline that involves the application of principles of physics for analysis, design, manufacturing, and maintenance of mechanical systems. Mechanical engineering is one of the oldest and broadest engineering disciplines.

It requires a solid understanding of core concepts including mechanics, kinematics, thermodynamics, fluid mechanics, and energy. Mechanical engineers use the core principles as well as other knowledge in the field to design and analyze motor vehicles, aircraft, heating and cooling systems, watercraft, manufacturing plants, industrial equipment and machinery, robotics, medical devices and more.

History

Applications of mechanical engineering are found in the records of many ancient and medieval societies throughout the globe. In ancient Greece, the works of Archimedes (287 BC–212 BC) and Heron of Alexandria (c. 10–70 AD) deeply influenced mechanics in the Western tradition. In China, Zhang Heng (78–139 AD) improved a water clock and invented a seismometer, and Ma Jun (200–265 AD) invented a chariot with differential gears. The medieval Chinese horologist and engineer Su Song (1020–1101 AD) incorporated an escapement mechanism into his astronomical clock tower two centuries before any escapement could be found in clocks of medieval Europe, as well as the world's first known endless power-transmitting chain drive.

During the years from 7th to 15th century, the era called the Islamic golden age, there have been remarkable contributions from Muslims in the field of mechanical technology, Al Jaziri, who was one of them, wrote his famous *Book of Knowledge of Ingenious Mechanical Devices* in 1206, which presented many mechanical designs. He is also considered to be the inventor of many of the mechanical devices which now form the very basic of mechanisms, such as crank and cam shafts.

During the early 19th century in England and Scotland, the development of machine tools led mechanical engineering to develop as a separate field within engineering, providing manufacturing machines and the engines to power them. The first British professional society of mechanical engineers was formed in 1847, thirty years after civil engineers formed the first such professional society. In the United States, the American Society of Mechanical Engineers (ASME) was formed in 1880, becoming the third such professional engineering society, after the American Society of Civil Engineers (1852) and the American Institute of Mining Engineers (1871). The first schools in the United States to offer an engineering education were the United States Military Academy in 1817, an institution now known as Norwich University in 1819, and Rensselaer Polytechnic Institute in 1825. Education in mechanical engineering has historically been based on a strong foundation in mathematics and science.

Technology

Mechanical engineering technology is the application of physical principles and current technological developments to the creation of useful machinery and operation design. Technologies such as solid models may be used as the basis for finite element analysis (FEA) and / or computational fluid dynamics (CFD) of the design. Through the application of computer-aided manufacturing (CAM), the models may also be used directly by software to create "instructions" for the manufacture of objects represented by the models, through computer numerically-controlled (CNC) machining or other automated processes, without the need for intermediate drawings.

Mechanical engineering technologists are also expected to understand and be able to apply concepts from the chemistry and electrical engineering fields. Mechanical engineering technologists are expected to apply current technologies and principals to machine and product design, production, and manufacturing processes.

Fundamental aspects of mechanical engineering technology include:

- Dynamics
- Statics
- Strength of materials
- Heat transfer
- Fluid mechanics/fluid dynamics
- Applied thermodynamics
- Machine design and kinematics
- Material science
- Manufacturing process
- Engineering drafting and standard familiarization classes
- Circuit and electrical analysis
- Instrumentation and measurement
- HVAC
- Hydraulics and pneumatics
- Quality assurance
- Technical communications

CHAPTER 1
BASIC CONCEPTS

Solid Mechanics

Mechanics is, in the most general sense, the study of forces and their effect upon matter. Typically, engineering mechanics is used to analyze and predict the acceleration and deformation (both elastic and plastic) of objects under known forces (also called loads) or stresses. Subdisciplines of mechanics include:

- Statics, the study of non-moving bodies under known loads
- Dynamics (or kinetics), the study of how forces affect moving bodies
- Mechanics of materials, the study of how different materials deform under various types of stress
- Fluid mechanics, the study of how fluids react to forces
- Continuum mechanics, a method of applying mechanics that assumes that objects are continuous (rather than discrete)

Mechanical engineers typically use mechanics in the design or analysis phases of engineering. If the engineering project were the design of a vehicle, statics might be employed to design the frame of the vehicle, in order to evaluate where the stresses will be most intense. Dynamics might be used when designing the car's engine, to evaluate the forces in the pistons and cams as the engine cycles. Mechanics of materials might be used to choose appropriate materials for the frame and engine. Fluid mechanics might be used to design a ventilation system for the vehicle (see HVAC), or to design the intake system for the engine.

Kinematics

Kinematics is the study of the motion of bodies (objects) and systems (groups of objects), while ignoring the forces that cause the motion. The movement of a crane and the oscillations of a piston in an engine are both simple kinematic systems. The crane is a type of open kinematic chain, while the piston is part of a closed four bar linkage.

Mechanical engineers typically use kinematics in the design and analysis of mechanisms. Kinematics can be used to find the possible range of motion for a given mechanism, or, working in reverse, can be used to design a mechanism that has a desired range of motion.

Mechatronics and Robotics

Mechatronics is an interdisciplinary branch of mechanical engineering, electrical engineering and software engineering that is concerned with integrating electrical and mechanical engineering to create hybrid systems. In this way, machines can be automated through the use of electric motors, servo-mechanisms, and other electrical systems in conjunction with special software. A common example of a mechatronics system is a CD-ROM drive. Mechanical systems open and close the drive, spin the CD and move the laser, while an optical system reads the data on the CD and converts it to bits. Integrated software controls the process and communicates the contents of the CD to the computer.

Robotics is the application of mechatronics to create robots, which are often used in industry to perform tasks that are dangerous, unpleasant, or repetitive. These robots may be of any shape and size, but all are preprogrammed and interact physically with the world. To create a robot, an engineer typically employs kinematics (to determine the robot's range of motion) and mechanics (to determine the stresses within the robot).

Robots are used extensively in industrial engineering. They allow businesses to save money on labor, perform tasks that are either too dangerous or too precise for humans to perform them economically, and to insure better quality. Many companies employ assembly lines of robots, and some factories are so robotized that they can run by themselves. Outside the factory, robots have been employed in bomb disposal, space exploration, and many other fields. Robots are also sold for various residential applications.

Structural Analysis and Failure Analysis

Structural analysis is the branch of mechanical engineering (and also civil engineering) devoted to examining why and how objects fail. Structural failures occur in two general modes: static failure, and fatigue failure. Static structural failure occurs when, upon being loaded (having a force applied) the object being analyzed either breaks or is deformed plastically, depending on the criterion for failure. Fatigue failure occurs when an object fails after a number of repeated loading and unloading cycles. Fatigue failure occurs because of imperfections in the object: a microscopic crack on the surface of the object, for instance, will grow slightly with each cycle (propagation) until the crack is large enough to cause ultimate failure.

Failure is not simply defined as when a part breaks, however; it is defined as when a part does not operate as intended. Some systems, such as the perforated top sections of some plastic bags, are designed to break. If these systems do not break, failure analysis might be employed to determine the cause.

Structural analysis is often used by mechanical engineers after a failure has occurred, or when designing to prevent failure. Engineers often use online documents and books such as those published by ASM[20] to aid them in determining the type of failure and possible causes.

Structural analysis may be used in the office when designing parts, in the field to analyze failed parts, or in laboratories where parts might undergo controlled failure tests.

Thermodynamics

Thermodynamics is an applied science used in several branches of engineering, including mechanical and chemical engineering. At its simplest, thermodynamics is the study of energy, its use and transformation through a system. Typically, engineering thermodynamics is concerned with changing energy from one form to another. As an example, automotive engines convert chemical energy (enthalpy) from the fuel into heat, and then into mechanical work that eventually turns the wheels.

Thermodynamics principles are used by mechanical engineers in the fields of heat transfer, thermofluids, and energy conversion. Mechanical engineers use thermo-science to design engines and power plants, heating, ventilation, and air-conditioning (HVAC) systems, heat exchangers, heat sinks, radiators, refrigeration, insulation, and others.

Drafting

Drafting or technical drawing is the means by which mechanical engineers create instructions for manufacturing parts. A technical drawing can be a computer model or hand-drawn schematic showing all the dimensions necessary to manufacture a part, as well as assembly notes, a list of required materials, and other pertinent information. A U.S. mechanical engineer or skilled worker who creates technical drawings may be referred to as a drafter or draftsman. Drafting has historically been a two-dimensional process, but computer-aided design (CAD) programs now allow the designer to create in three dimensions.

Instructions for manufacturing a part must be fed to the necessary machinery, either manually, through programmed instructions, or through the use of a computer-aided manufacturing (CAM) or combined CAD/CAM program. Optionally, an engineer may also manually manufacture a part using the technical drawings, but this is becoming an increasing rarity, with the advent of computer numerically controlled (CNC) manufacturing. Engineers primarily manually manufacture parts in the areas of applied spray coatings, finishes, and other processes that cannot economically or practically be done by a machine.

Drafting is used in nearly every subdiscipline of mechanical engineering, and by many other branches of engineering and architecture. Three-dimensional models created using CAD software are also commonly used in finite element analysis (FEA) and computational fluid dynamics (CFD).

CHAPTER 2
ENGINEERING THERMODYNAMICS

Engineers use thermodynamics to calculate the fuel efficiency of engines, and to find ways to make more efficient systems, be they rockets, refineries, or nuclear reactors. Thermodynamics is the science that deals with transfer of heat and work. Engineering thermodynamics develops the theory and techniques required to use empirical thermodynamic data effectively. However, with the advent of computers most of these techniques are transparent to the engineer, and instead of looking data up in tables, computer applications can be queried to retrieve the required values and use them in calculations. There are even applications which are tailored to specific areas which will give answers for common design situations. But thorough understanding will only come with knowledge of underlying principles, and the ability to judge the limitations of empirical data is perhaps the most important gain from such knowledge.

Thermodynamics is the study of the relationships between HEAT (thermos) and WORK (dynamics). Thus, it deals with energy interactions in physical systems. Classical thermodynamics can be stated in four laws called the zeroth, first, second, and third laws respectively. The laws of thermodynamics are empirical, i.e., they are deduced from experience, and supported by a large body of experimental evidence.

Introduction to Classical Thermodynamics

Thermodynamics is the study of energies. More specifically, introductory thermodynamics is the study of energy transfer in systems. Classical thermodynamics consists of methods and constructs that are used to "account" for macroscopic energy transfer. In fact, energy accounting is an appropriate synonym for classical thermodynamics. In much the same way that accountants balance money in and money out of a bank account, rocket scientists simply balance the energy in and out of a rocket engine. Of course just as a bank account's balance is obfuscated by arcane devices such as interest rates and currency exchange, so too is thermodynamics clouded with seemingly difficult concepts such as irreversibility and enthalpy. But, also just like accounting, a careful review of the rules suggests a coherent strategy for maintaining tabs on a particular account.

If a statement about the simplicity of thermodynamics failed to convert would-be students, they may be captured with a few words on the importance of understanding energy transfer in our society. Up until about 150 years ago or so, the earth's economy was primarily fueled by carbohydrates. That is to say, humans got stuff done by converting food, through a biological process, to fuel we could spend to do work (e.g. raise barns). This was a hindrance to getting things accomplished because, as it turned out, most individuals had to use the brunt of that energy to grow and cultivate more carbohydrates (e.g. crops and livestock). We won't even talk about how much food the horses ate!

Today, we have the luxury, primarily through an understanding of energy, to concentrate our energy production into efficient low maintenance operations. Massive power plants transfer energy to power tools for raising barns. Extremely efficient rocket engines tame and direct massive amounts of energy to blast TV satellites into orbit. This improvement in energy mastery frees humanity's time to engage in more worthwhile activities such as watching cable TV. Although most are content to blissfully ignore the intricacies that command their way of life, I challenge you to embrace the contrary.

By no means is the energy battle over. Understanding energy transfer and energy systems is only one step toward destroying the limits to what humanity can next accomplish.

The Main Macroscopic Forms of Energy

There are three primary forms of energy that are discussed in macroscopic thermodynamics. Several other forms of energy exist, but they generally exist on a microscopic level and should be deferred to more advanced study.

Kinetic Energy

The first form (probably most easily understood idea of energy) is defined by the motion of an object. Kinetic energy is the energy of a moving mass. For instance, a moving car will have more kinetic energy than a stationary car. The same car traveling at 60 km/h has more kinetic energy than it does traveling at 30 km/h.

Kinetic Energy = (1/2) x (mass) x (velocity)2

Ratio of v602 / v302 =((60)(1,000 meters/sec)(3600 sec/hour))2 / ((30)(1,000 meters/sec)(3,600 sec/hour))2

So, once the algebra is completed properly we find the vehicle traveling at 60 km/h has four times the kinetic energy as when it is traveling at 30 km/h, while the vehicle has zero kinetic energy when it is stationary because the velocity = zero results in 1/2mv2 being zero.

Potential Energy

The second type of energy is called potential energy. Gravitational potential energy describes the energy due to elevation. A car at a height of 50 m has more potential energy than a car at a height of 25 m. This may be understood more easily if the car is allowed to drop from its height. On impact with the earth at 0 m, the car that initially rested at 50 m will have more kinetic energy because it was moving faster (allowed more time to accelerate). The idea that potential energy can convert to kinetic energy is the first idea of energy transfer. Transfers between kinetic and potential energy represent one type of account balance rocket scientists need be aware of.

Internal Energy of Matter

The third and most important concept of energy is reflected by temperature. The internal energy of matter is measured by its temperature. Hot water has more internal energy than the same amount of cold water. Internal energy is a measure of kinetic energy of the molecules and atoms that make up the substance. Since each atom or molecule is acting on its own accord, this internal energy is different from the bulk kinetic energy associated with the movement of the entire solid. The internal energy of matter is exhibited by molecular motion. The molecules of a gas at high temperature zip around their container constantly colliding with walls and other molecules. The molecules of a high temperature solid also move around a lot; however, since they are stuck together with other molecules, the most they can do is vibrate in place.

In a nutshell, the above forms of energy are studied in classical thermodynamics. Those forms of energy are allowed to transfer among each other as well as in to or out of a system. Thermodynamics essentially provides some definitions for interpreting thermodynamic systems. It then goes on to define an important rule about fairly balancing energy and one rule about the quality of energy. (Some energy is more valuable.) Understanding the framework and the few rules that govern macroscopic thermodynamics proves to be an incredibly powerful set of tools for analyzing a myriad of not only engineering problems, but issues of practical concern.

Properties of Pure Substances

A cursory review of properties will introduce the variables of thermodynamics to the student. Properties of substances are things such as mass, temperature, volume, and pressure. Properties are used to define the current state of a substance. Several more properties exist to describe substances in thermodynamics, but a stronger understanding of theory is required for their definition and application.

Properties can be intensive, if they are point properties (properties that make sense for a point) or extensive, if they depend on the amount of matter in the system. Examples of extensive properties of systems are mass of system, number of moles of a substance in a system, and overall or total volume of a system. These properties depend on how much matter of the system you measure. Examples of intensive properties are pressure, temperature, density, volume per mass, molar volume (which is volume per mole), and average molecular weight (or molecular mass). These properties are the same regardless of how you vary the amount of mass of the substance.

Properties are like the variables for substances in that their values are all related by an equation. The relationship between properties is expressed in the form of an equation which is called an equation of state. Perhaps the most famous state equation is the Ideal Gas Law. The ideal gas law relates the pressure, volume, and temperature of an ideal gas to one another.

Volume

The SI unit for volume is m3. Volume is an extensive property, but both volume per mass and molar volume are intensive properties since they do not depend on the measured mass of the system. A process during which the volume of the system remains constant is called an isochoric process.

Pressure

The SI unit for pressure is Pa (Pascal), which is equivalent to a N / (m2). Pressure is an intensive property. A process in which pressure remains constant is called isobaric.

Temperature

The concept of temperature is fundamental to thermodynamics. We know that a body at high temperature will transfer energy to one at lower temperature. Consider two bodies with different temperatures in contact with each other. Net energy transfer will be from the hotter body to the colder body. At some point, the net energy transfer will be zero, and the bodies are said to be in thermal equilibrium. Bodies in thermal equilibrium are defined to have the same temperature.

Thermodynamic Systems

In general, a system is a collection of objects, and there is a lot of subtlety in the way it is defined, as in set theory. However, in thermodynamics, it is a much more straightforward concept. A thermodynamic system is defined as a volume in space or a well defined set of materials (matter). The imaginary outer edge of the system is called its boundary.

As can be seen from the definition, the boundary can be fixed or moving. A system in which matter crosses the boundary is called an open system. Similarly, a system in which no matter enters or leaves (i.e. crosses the boundary) is called a closed system.

Figure 1 shows a piston cylinder arrangement, where a gas is compressed by the piston. As can be seen, due to an opening in the cylinder, gas can escape outside as the piston moves inwards, and gas enters the system when the piston moves outwards. Thus, it is an open system.

Now consider a similar system, but one in which gas cannot escape. An example is the Sterling engine shown in Figure 2. Thus the model of this configuration is a closed system.

The region outside the system is called the surroundings. The system and the surroundings together are called the Universe. A system which does not exchange matter or energy with the surroundings is called an isolated system.

Another term sometimes used instead of system is control volume. In the case of a closed system, in which the mass of matter inside the system remains constant, the control volume is referred to as control mass. A control volume is said to be enclosed by a control surface.

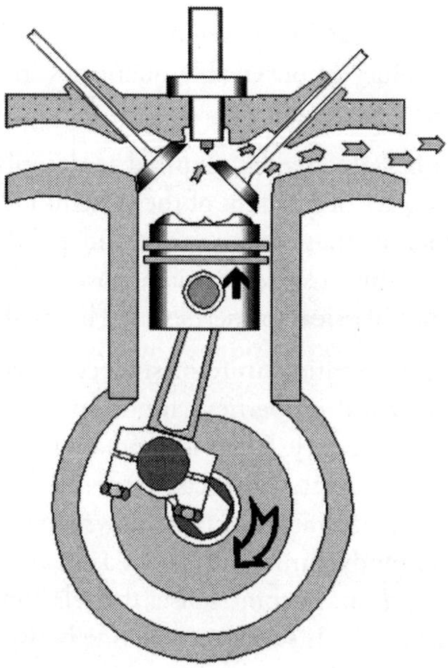

Figure 1.

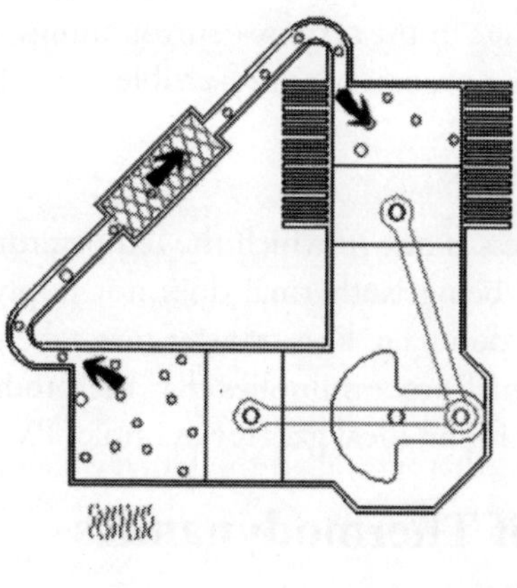

Figure 2.

Systems

Classical thermodynamics deals with systems in equilibrium. The equilibrium state is defined by the values of observable quantities in the system. These are called system properties.

The minimum number of variables required to describe the system depends on the complexity or degrees of freedom of the system. Degrees of freedom refer to the number of properties that can be varied independently of each other in a system. Some of the common system variables are pressure, temperature, and density, though any other physical properties may be used.

Consistent with the axiomatic nature of subject development, many of the relationships between physical properties cannot be completely specified without further development of theory. What is good about classical thermodynamics is that many of the axioms stated here can be derived using techniques of statistical thermodynamics. And statistical thermodynamics gives results in many cases where classical thermodynamics fails, such as in the specific heats of gases with many degrees of freedom. In some sense, the relationship between classical and statistical thermodynamics is similar to the one between classical and quantum mechanics, i.e., classical thermodynamics approximates statistical thermodynamics in the macroscopic limit.

Processes

A change in the system state is called a process. When the initial and final states of a process are the same, the process is called a cycle. If a process can be run in reverse with no change in the system + surroundings, then the process is called a reversible process. If a process is not reversible it is called an irreversible process.

Isothermal Process

An isothermal process is one in which the temperature remains constant. Please note that a process being isothermal does not imply anything about the heat transferred or work done, i.e. heat transfer may take place during an isothermal process. An isothermal process implies that the product of the volume and the pressure is constant for an ideal gas (for example, $PV = \text{Constant}$).

Zeroth Law of Thermodynamics

If a system A is in thermal equilibrium with another system B and also with a third system C, then all of the systems are in thermal equilibrium with each other. This is called the zeroth law of thermodynamics. This is how a thermometer works. If a thermometer is placed in a substance for temperature measurement, the thermometer's glass comes into thermal equilibrium with the substance. The glass then comes into thermal equilibrium with the liquid (mercury, alcohol, etc . . .) inside the thermometer. Because the substance is in thermal equilibrium with the glass and the glass is in thermal equilibrium with the inner liquid, the substance and liquid must be in thermal equilibrium by the zeroth law. And because they are thermally equivalent, they must have the same temperature.

Temperature Measurement

Temperature is measured by observing some property of the system which varies with temperature. Such a property is called thermometric property.

- The volumes of most liquids increase with temperature.
- The length of a metal rod increases as the temperature increases.
- The pressure of a constant volume of gas increases with temperature

It is useful to establish a temperature scale so that a cardinal relationship can be established between various systems at different temperatures. This is done by defining the temperature t as a function of a thermometric property X, such that the temperature is a linear function of X, i.e., equal changes in the property X give rise to equal changes in the temperature. Such a linear function is $t = a + b X$, for which one needs to assign arbitrary temperatures to two values of X to find the values of the constants a and b.

For example, in the case of the Celsius scale, the measurements are based on properties of water at the boiling point and melting point. Suppose the value of the thermometric property is Xb for the normal boiling point and Xm for the normal melting point. Then the temperature is given by $t = 100 (X - Xm)/(Xb - Xm)$, where X is the thermometric property at temperature t, and we have chosen tm = 0°C and tb = 100°C. The normal melting and boiling points are the temperatures of melting and boiling at 1 atmosphere pressure.

The major temperature scales are the Celsius scale (°C), the Fahrenheit (°F) and the Kelvin (K) scale. Note the absence of the ° sign for kelvin—it is not degrees Kelvin, but kelvins, not capitalized when spelled out, and with the normal English plural which was added to the degree when Kelvin was an adjective modifying that unit.

Different thermometers are used for different temperature ranges. As the reader might have guessed by now, this means that the different thermometers will only agree on the fixed points. However, a set of thermometers have been carefully selected and calibrated so that this is not a big issue in practice.

The standard in this case is the International Temperature Scale, which was introduced in 1927, and revised in 1948, 1968, and 1990. The latest scale is denoted by T90 for the Kelvin scale and is defined from 0.65 K upwards. For instance, between 0.65 K and 5.0 K T90 is defined in terms of the vapor-pressure temperature relations of 3He and 4He. The ranges for different materials overlap and any of the valid materials can be used as a standard in the overlapping region.

The Ideal Gas

Recall that a thermodynamic system may have a certain substance or material whose quantity can be expressed in mass or moles in an overall volume. These are extensive properties of the system. If the substance is evenly distributed throughout the volume in question, then a value of volume per amount of substance may be used as an intensive property. For an example, for an amount called a mole, volume per mole is typically called molar volume. Also, a volume per mass for a specific substance may be called specific volume. In such cases, an equation of state may relate the three intensive properties, temperature, pressure, and molar or specific volume.

A simple but very useful equation of state is for an ideal gas. The ideal gas is a useful notion in thermodynamics, as it is a simple system that depends on two independent properties. An ideal gas is one that has no intermolecular interactions except for completely elastic collisions with other molecules. For a closed system containing an ideal gas, the state can be specified by giving the values of any two of pressure, temperature, and molar volume.

Consider a system, an ideal gas enclosed in a container. Starting from an initial state 1, where the temperature is T1, its temperature is changed to T2 through a constant pressure process and then a constant molar volume process, then the ratio of pressures is found to be the same as the ratio of molar volumes. Suppose the initial value of the pressure and molar volume are p1 and V1 respectively, and final value of pressure and molar volume are p2 and V2 respectively. Note that we haven't chosen a specific scale for the temperature (like say, the Celsius scale). Now, suppose we were to choose a scale such that T1/T2 = p1/p2, we can show that the value of pV/T is constant for an ideal gas, so that it obeys the gas equation pV = RT, where p is the absolute pressure, V is the molar volume, and R is a constant known as the universal gas constant. The temperature T is the absolute temperature in the ideal gas scale, and the scale is found to be the same as the thermodynamic temperature scale. The thermodynamic temperature scale will be defined after the statement of the second law of thermodynamics.

This equation $pV = RT$ is called the equation of state for an ideal gas, and is known as the ideal gas equation. Most common gases obey the ideal gas equation unless they are compressed or cooled to extreme states, so this is a very useful relation. A similar equation may be written where, for the specific type of gas, specific volume is used instead of molar volume and a specific gas constant is used instead of the universal gas constant. This then writes as $pv = mrT$.

From experiment, it is easy to show that the thermodynamic temperature T is related to the Celsius temperature t by the equation: $T = t + 273.15$. The zero of thermodynamic temperature scale is 0 K, and its significance will be clear when we discuss the second law of thermodynamics.

CHAPTER 3
A MECHANICAL ENGINEER LOOKS AT THE AUTOMATIC TRANSMISSION

If you have ever driven a car with an automatic transmission, then you know that there are two big differences between an automatic transmission and a manual transmission:

- There is no clutch pedal in an automatic transmission car.
- There is no gear shift in an automatic transmission car. Once you put the transmission into Drive, everything else is automatic!

Both the automatic transmission (plus its torque converter) and a manual transmission (with its clutch) accomplish exactly the same thing, but they do it in totally different ways. It turns out that the way an automatic transmission does it is absolutely amazing! An example of an automatic transmission is shown in figure 1.

In this chapter, we'll work our way through an automatic transmission. We'll start with the key to the whole system: planetary gearsets. Then we'll see how the transmission is put together, learn how the controls work and discuss some of the intricacies involved in controlling a transmission.

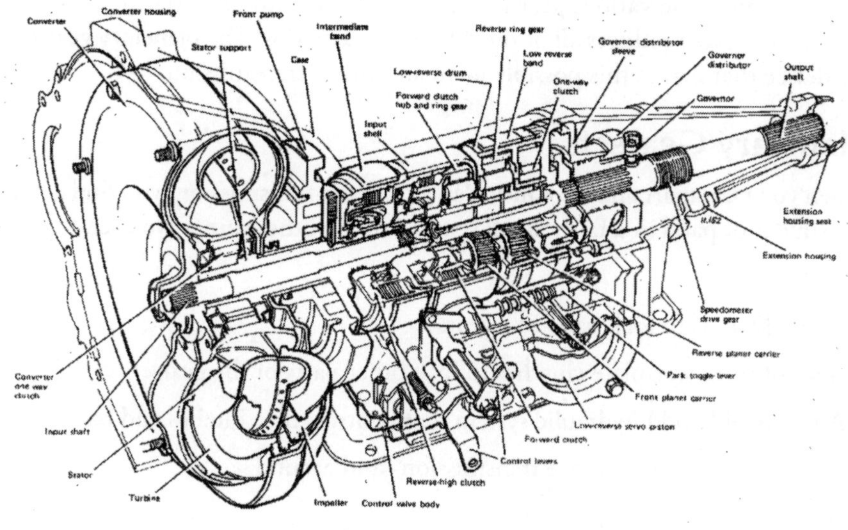

The Ranger Station - Exploded View Of C-4 Transmission

Figure 1.

Some Basics

Just like that of a manual transmission, the automatic transmission's primary job is to allow the engine to operate in its narrow range of speeds while providing a wide range of output speeds.

Without a transmission, cars would be limited to one gear ratio, and that ratio would have to be selected to allow the car to travel at the desired top speed. If you wanted a top speed of 80 mph, then the gear ratio would be similar to third gear in most manual transmission cars.

You've probably never tried driving a manual transmission car using only third gear. If you did, you'd quickly find out that you had almost no acceleration when starting out, and at high speeds, the engine would be screaming along near the red-line. A car like this would wear out very quickly and would be nearly undriveable.

So the transmission uses gears to make more effective use of the engine's torque, and to keep the engine operating at an appropriate speed.

The key difference between a manual and an automatic transmission is that the manual transmission locks and unlocks different sets of gears to the output shaft to achieve the various gear ratios, while in an automatic transmission, the same set of gears produces all of the different gear ratios. The planetary gearset is the device that makes this possible in an automatic transmission.

Planetary Gearsets

When you take apart and look inside an automatic transmission, you find a huge assortment of parts in a fairly small space. Among other things, you see:

- A set of bands to lock parts of a gearset
- A set of three wet-plate clutches to lock other parts of the gearset
- An incredibly odd hydraulic system that controls the clutches and bands
- A large gear pump to move transmission fluid around

The center of attention is the planetary gearset. An example of one is shown below:

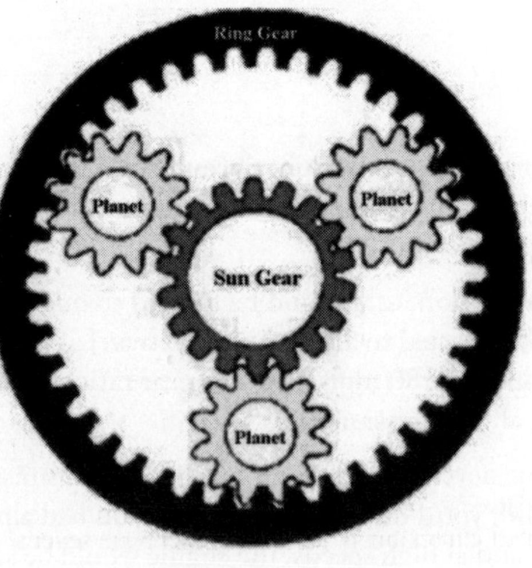

About the size of a cantaloupe, this one part creates all of the different gear ratios that the transmission can produce. Everything else in the transmission is there to help the planetary gearset do its thing. An automatic transmission contains two complete planetary gearsets folded together into one component.

Any planetary gearset has three main components:

- The sun gear
- The planet gears and the planet gears' carrier
- The ring gear

Each of these three components can be the input, the output or can be held stationary. Choosing which piece plays which role determines the gear ratio for the gearset. Let's take a look at a single planetary gearset.

One of the planetary gearsets from our transmission has a ring gear with 72 teeth and a sun gear with 30 teeth. We can get lots of different gear ratios out of this gearset.

Input	Output	Stationary	Gear Ratio
A Sun (S)	Planet Carrier (C)	Ring (R)	1 + R/S 3.4:1
B Planet Carrier (C)	Ring (R)	Sun (S)	1 / (1 + S/R) 0.71:1
C Sun (S)	Ring (R)	Planet Carrier (C)	-R/S -2.4:1

Also, locking any two of the three components together will lock up the whole device at a 1:1 gear reduction. Notice that the first gear ratio listed above is a reduction—the output speed is slower than the input speed. The second is an overdrive—the output speed is faster than the input speed. The last is a reduction again, but the output direction is reversed. There are several other ratios that can be gotten out of this planetary gear set, but these are the ones that are relevant to our automatic transmission.

So this one set of gears can produce all of these different gear ratios without having to engage or disengage any other gears. With two of these gearsets in a row, we can get the four forward gears and one reverse gear our transmission needs. We'll put the two sets of gears together in the next section.

Gears

This automatic transmission uses a set of gears, called a compound planetary gearset, that looks like a single planetary gearset but actually behaves like two planetary gearsets combined. It has one ring gear that is always the output of the transmission, but it has two sun gears and two sets of planets.

First Gear

In first gear, the smaller sun gear is driven clockwise by the turbine in the torque converter. The planet carrier tries to spin counterclockwise, but is held still by the one-way clutch (which only allows rotation in the clockwise direction) and the ring gear turns the output. The small gear has 30 teeth and the ring gear has 72, so referring to the chart given earlier, the gear ratio is:

Ratio = -R/S = -72/30 = -2.4:1

So the rotation is negative 2.4:1, which means that the output direction would be opposite the input direction. But the output direction is really the same as the input direction—this is where the trick with the two sets of planets comes in. The first set of planets engages the second set, and the second set turns the ring gear; this combination reverses the direction. You can see that this would also cause the bigger sun gear to spin; but because that clutch is released, the bigger sun gear is free to spin in the opposite direction of the turbine (counter-clockwise).

Second Gear

This transmission does something really neat in order to get the ratio needed for second gear. It acts like two planetary gearsets connected to each other with a common planet carrier.

The first stage of the planet carrier actually uses the larger sun gear as the ring gear. So the first stage consists of the sun (the smaller sun gear), the planet carrier, and the ring (the larger sun gear).

The input is the small sun gear; the ring gear (large sun gear) is held stationary by the band, and the output is the planet carrier. For this stage, with the sun as input, planet carrier as output, and the ring gear fixed, the formula is:

1 + R/S = 1 + 36/30 = 2.2:1

The planet carrier turns 2.2 times for each rotation of the sun gear. At the second stage, the planet carrier acts as the input for the second planetary gear set, the larger sun gear (which is held stationary) acts as the sun, and the ring gear acts as the output, so the gear ratio is:

1 / (1 + S/R) = 1 / (1 + 36/72) = 0.67:1

To get the overall reduction for second gear, we multiply the first stage by the second, 2.2 x 0.67, to get a 1.47:1 reduction. This may sound wacky, but it works.

Third Gear

Most automatic transmissions have a 1:1 ratio in third gear. You'll remember from the previous section that all we have to do to get a 1:1 output is lock together any two of the three parts of the planetary gear. With the arrangement in this gearset it is even easier—all we have to do is engage the clutches that lock each of the sun gears to the turbine.

If both sun gears turn in the same direction, the planet gears lockup because they can only spin in opposite directions. This locks the ring gear to the planets and causes everything to spin as a unit, producing a 1:1 ratio.

Overdrive

By definition, an overdrive has a faster output speed than input speed. It's a speed increase—the opposite of a reduction. In this transmission, engaging the overdrive accomplishes two things at once. In order to improve efficiency, some cars have a mechanism that locks up the torque converter so that the output of the engine goes straight to the transmission.

In this transmission, when overdrive is engaged, a shaft that is attached to the housing of the torque converter (which is bolted to the flywheel of the engine) is connected by clutch to the planet carrier. The small sun gear freewheels, and the larger sun gear is held by the overdrive band. Nothing is connected to the turbine; the only input comes from the converter housing. Let's go back to our chart again, this time with the planet carrier for input, the sun gear fixed and the ring gear for output.

Ratio = 1 / (1 + S/R) = 1 / (1 + 36/72) = 0.67:1

So the output spins once for every two-thirds of a rotation of the engine. If the engine is turning at 2000 rotations per minute (RPM), the output speed is 3000 RPM. This allows cars to drive at freeway speed while the engine speed stays nice and slow.

Reverse

Reverse is very similar to first gear, except that instead of the small sun gear being driven by the torque converter turbine, the bigger sun gear is driven, and the small one freewheels in the opposite direction. The planet carrier is held by the reverse band to the housing. So, according to our equations from the last page, we have:

Ratio = -R/S = 72/36 = 2.0:1

So the ratio in reverse is a little less than first gear in this transmission.

Gear Ratios

This transmission has four forward gears and one reverse gear. Let's summarize the gear ratios, inputs and outputs:

Gear	Input	Output	Fixed Gear Ratio
1st	30-tooth sun	72-tooth ring Planet carrier	2.4:1
2nd	30-tooth sun	Planet carrier 36-tooth ring	2.2:1
3rd	30- and 36-tooth suns	72-tooth ring	1.0:1
OD	Planet carrier 72-tooth ring	36-tooth sun	0.67:1
Rev.	36-tooth sun	72-tooth ring Planet carrier	-2.0:1

After reading this section, you are probably wondering how the different inputs get connected and disconnected. This is done by a series of clutches and bands inside the transmission. In the next section, we'll see how these work.

Clutches and Bands

In this transmission, when overdrive is engaged, a shaft that is attached to the housing of the torque converter (which is bolted to the flywheel of the engine) is connected by clutch to the planet carrier. The small sun gear freewheels, and the larger sun gear is held by the overdrive band. Nothing is connected to the turbine; the only input comes from the converter housing.

To get the transmission into overdrive, lots of things have to be connected and disconnected by clutches and bands. The planet carrier gets connected to the torque converter housing by a clutch. The small sun gets disconnected from the turbine by a clutch so that it can freewheel. The big sun gear is held to the housing by a band so that it could not rotate. Each gear shift triggers a series of events like these, with different clutches and bands engaging and disengaging. Let's take a look at a band.

Bands

The bands in a transmission are, literally, steel bands that wrap around sections of the gear train and connect to the housing. They are actuated by hydraulic cylinders inside the case of the transmission. One of the bands is shown below.

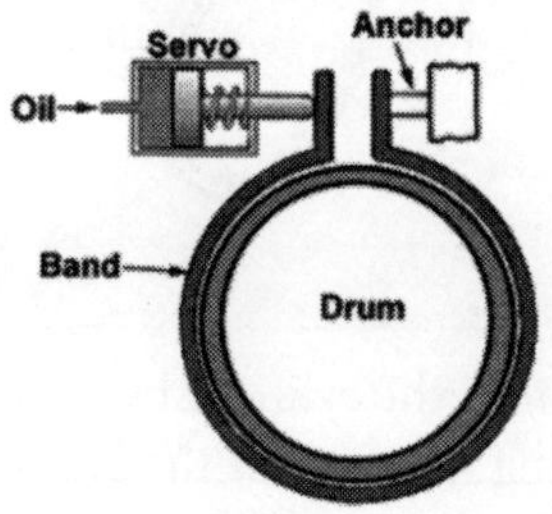

In the previous figure, you can see one of the bands in the housing of the transmission. The gear train is removed. The metal rod is connected to the piston, which actuates the band.

The pistons that actuate the bands are visible here.

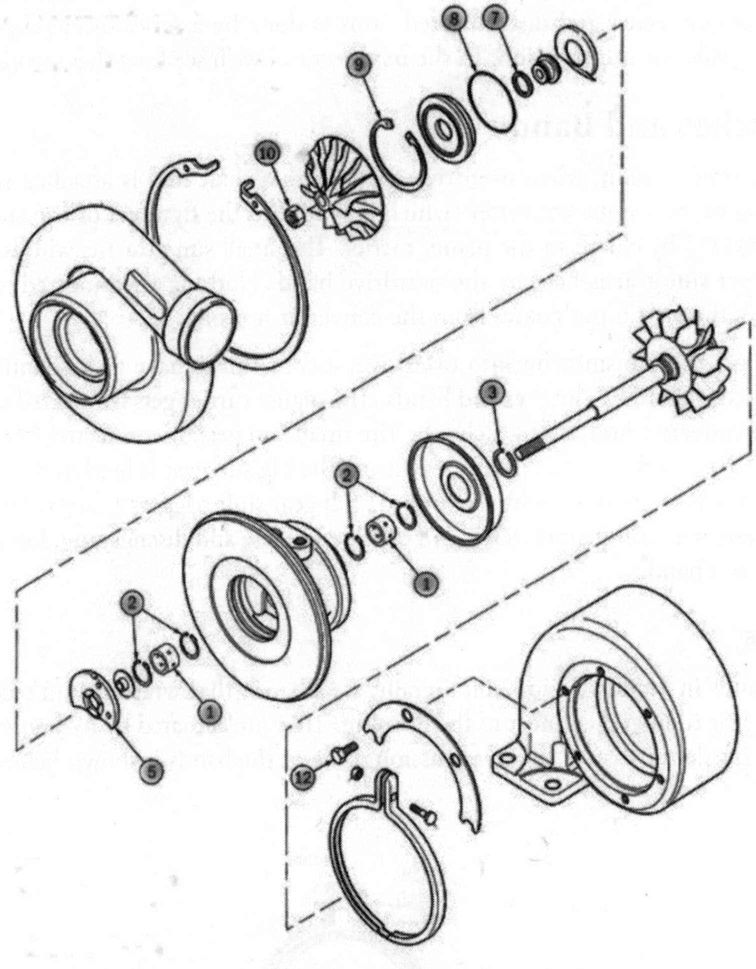

Hydraulic pressure, routed into the cylinder by a set of valves, causes the pistons to push on the bands, locking that part of the gear train to the housing.

Clutches

The clutches in the transmission are a little more complex. Each clutch is actuated by pressurized hydraulic fluid that enters a piston inside the clutch. Springs make sure that the clutch releases when the pressure is reduced. Below you can see the piston and the clutch drum.

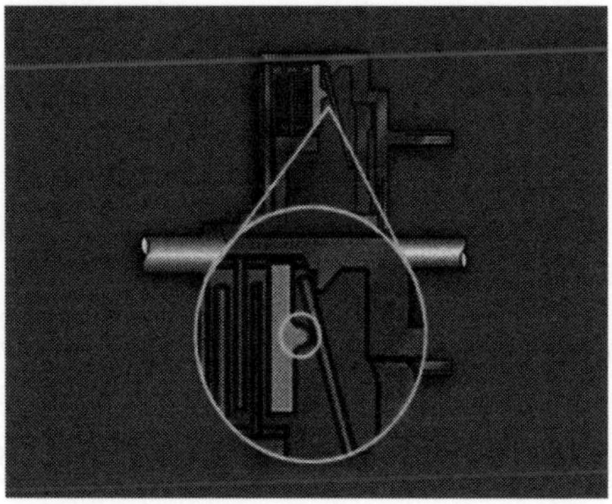

The next figure shows the alternating layers of clutch friction material and steel plates. The friction material is splined on the inside, where it locks to one of the gears. The steel plate is splined on the outside, where it locks to the clutch housing. These clutch plates are replaced when the transmission is rebuilt.

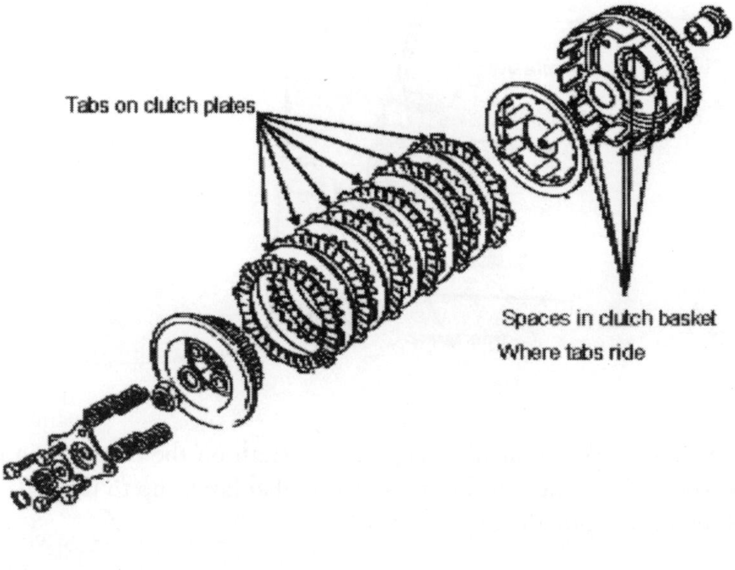

The pressure for the clutches is fed through passageways in the shafts. The hydraulic system controls which clutches and bands are energized at any given moment.

When You Put the Car in Park

It may seem like a simple thing to lock the transmission and keep it from spinning; but there are actually some complex requirements for this mechanism:

- You have to be able to disengage it when the car is on a hill (the weight of the car is resting on the mechanism).
- You have to be able to engage the mechanism even if the lever does not line up with the gear.
- Once engaged, something has to prevent the lever from popping up and disengaging.

The mechanism that does all this is pretty neat. Let's look at some of the parts first.

The output of the transmission is shown below. The notches are engaged by the parking-brake mechanism to hold the car still.

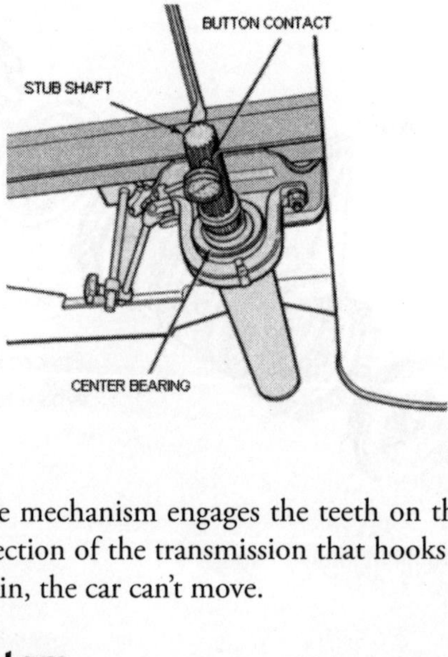

The parking-brake mechanism engages the teeth on the output to hold the car still. This is the section of the transmission that hooks up to the drive shaft, so if this part can't spin, the car can't move.

Hydraulic System

The automatic transmission in your car has to do numerous tasks. You may not realize how many different ways it operates. For instance, here are some of the features of an automatic transmission:

- If the car is in overdrive (on a four-speed transmission), the transmission will automatically select the gear based on vehicle speed and throttle pedal position.
- If you accelerate gently, shifts will occur at lower speeds than if you accelerate at full throttle.
- If you floor the gas pedal, the transmission will downshift to the next lower gear.
- If you move the shift selector to a lower gear, the transmission will downshift unless the car is going too fast for that gear. If the car is going too fast, it will wait until the car slows down and then downshift.
- If you put the transmission in second gear, it will never downshift or upshift out of second, even from a complete stop, unless you move the shift lever.

The Pump

Automatic transmissions have a neat pump, called a gear pump (shown below). The pump is usually located in the cover of the transmission. It draws fluid from a sump in the bottom of the transmission and feeds it to the hydraulic system. It also feeds the transmission cooler and the torque converter.

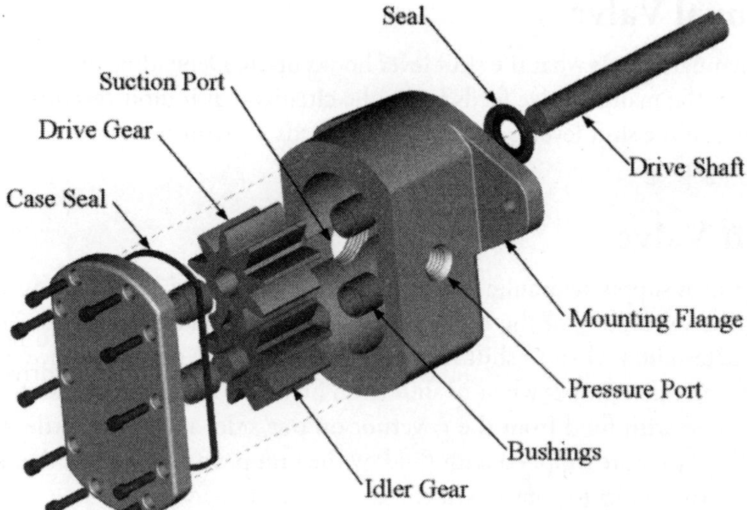

The inner gear of the pump hooks up to the housing of the torque converter, so it spins at the same speed as the engine. The outer gear is turned by the inner gear, and as the gears rotate, fluid is drawn up from the sump on one side of the crescent and forced out into the hydraulic system on the other side.

The Governor

The governor is a clever valve that tells the transmission how fast the car is going. It is connected to the output, so the faster the car moves, the faster the governor spins. Inside the governor is a spring-loaded valve that opens in proportion to how fast the governor is spinning—the faster the governor spins, the more the valve opens. Fluid from the pump is fed to the governor through the output shaft.

Throttle Valve or Modulator

To shift properly, the automatic transmission has to know how hard the engine is working. There are two different ways that this is done. Some cars have a simple cable linkage connected to a throttle valve in the transmission. The further the gas pedal is pressed, the more pressure is put on the throttle valve. Other cars use a vacuum modulator to apply pressure to the throttle valve. The modulator senses the manifold pressure, which drops when the engine is under a greater load.

Manual Valve

The manual valve is what the shift lever hooks up to. Depending on which gear is selected, the manual valve feeds hydraulic circuits that inhibit certain gears. For instance, if the shift lever is in third gear, it feeds a circuit that prevents overdrive from engaging.

Shift Valves

Shift valves supply hydraulic pressure to the clutches and bands to engage each gear. The valve body of the transmission contains several shift valves. The shift valve determines when to shift from one gear to the next. For instance, the 1 to 2 shift valve determines when to shift from first to second gear. The shift valve is pressurized with fluid from the governor on one side, and the throttle valve on the other. They are supplied with fluid by the pump, and they route that fluid to one of two circuits to control which gear the car runs in.

As car speed increases, the pressure from the governor builds. This forces the shift valve over until the first gear circuit is closed, and the second gear circuit opens. Since the car is accelerating at light throttle, the throttle valve does not apply much pressure against the shift valve.

When the car accelerates quickly, the throttle valve applies more pressure against the shift valve. This means that the pressure from the governor has to be higher (and therefore the vehicle speed has to be faster) before the shift valve moves over far enough to engage second gear.

Each shift valve responds to a particular pressure range; so when the car is going faster, the 2-to-3 shift valve will take over, because the pressure from the governor is high enough to trigger that valve.

Electronic Controls

Electronically controlled transmissions, which appear on some newer cars, still use hydraulics to actuate the clutches and bands, but each hydraulic circuit is controlled by an electric solenoid. This simplifies the plumbing on the transmission and allows for more advanced control schemes.

In the last section we saw some of the control strategies that mechanically controlled transmissions use. Electronically controlled transmissions have even more elaborate control schemes. In addition to monitoring vehicle speed and throttle position, the transmission controller can monitor the engine speed, if the brake pedal is being pressed, and even the anti-lock braking system.

Using this information and an advanced control strategy based on fuzzy logic—a method of programming control systems using human-type reasoning—electronically controlled transmissions can do things like:

- Downshift automatically when going downhill to control speed and reduce wear on the brakes
- Upshift when braking on a slippery surface to reduce the braking torque applied by the engine
- Inhibit the upshift when going into a turn on a winding road.

CHAPTER 4
AERODYNAMICS AND RACING CARS

By studying the trends of the last thirty years we see that aerodynamics is a reason behind the success of a racing car, since tire and engine technology have made a comparatively small progress. Furthermore, tires are not manufactured to fit the particular needs of a single racing team, while engines often are developed in series for more than one team. This conclusion leads us to the analysis of the aerodynamic tools: computational aerodynamics (CAD), wind tunnel testing and aerodynamic design and optimisation.

The Role of the Wind Tunnel

Racing teams have been devoting more and more time to the development of the aerodynamics of their cars. They have done so with track and wind tunnel testing. Track testing is widely recognised for being too expensive and dependent on many casual events. The advantage is of course that the car is tested in its actual configuration in a real world situation.

The wind tunnel is the technical answer of the aerodynamic engineers. The wind tunnels are nowadays very sophisticated, and allow a wide range of studies including modelling of the car in compete configuration, ground plane simulation, etc.

Computer-Aided Design

Recent progress in software and computer hardware has made available sophisticated methods for real time simulations of full configurations. Integrated CAD-grid generation- and finite element systems have placed the design to a very rational level from a cut-and-try approach followed for many years. The computer approach is probably only limited by the hardware power, although there are still enormous difficulties in treating a large class of problems.

Aerodynamic Design

Aerodynamic design may as well be the largest and the oldest problem in the already vast field of aerodynamics.

The problem, in fact, consists in designing aerodynamic systems and/ or any of their components, propulsion systems or any or their components to perform specified operations in a range of design points.

The field spans all the aerodynamic knowledge, and stretches into design as considered a separate engineering science.

Design techniques have not evolved farther than the rule-of-thumb until the development of the wind tunnel. The experimental work has produced classes of airfoils widely used by the aeronautic industry for some thirty years.

Downforce

Since the introduction of aerodynamics, it has been clear that the proper distribution of downforce has the most evident impact on car performances. Racing regulations are strongly focused on chassis characteristics, and give strict limits to size, position, and type of devices allowed and prohibited. Nevertheless, cornering speeds have reached the 4g.

However, the car also needs to be able to go around corners and this presents a differing set of aerodynamic needs, namely downforce, which is the force harnessed from the airflow over a car that presses it down onto the track. Downforce is necessary in maintaining high speeds through the corners and forces the car to the track. Downforce is necessary for maintaining speed through the corners.

Last but not least, downforce gave cars superior traction, and thus made them less likely to spin when being driven around a corner, enabling them to turn at greater speed.

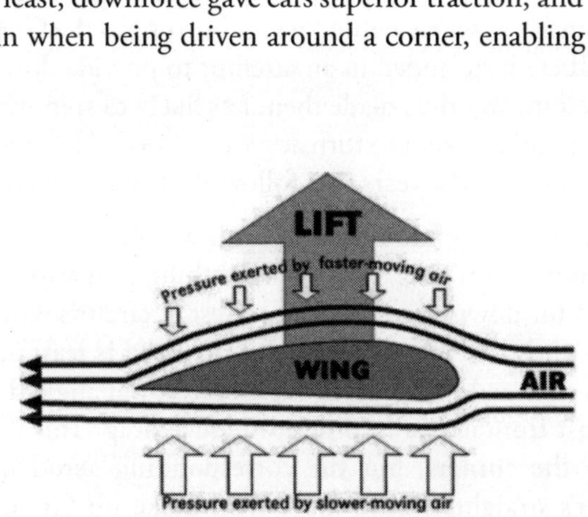

Downforce in an airplane wing is based on the same aerodynamic principle as that in race cars.

Ground Effect

The ground effect car was introduced by Team Lotus in their 1978 Formula 1 car. With the introduction of the ground effect a few years later the vehicles used a third element (the underbody) to produce downforce, and hence improve the improvements. By shaping the underbody with appropriate channels, and using side pods to increase the effective area, the car provided much larger values of downforce. The idea, again, was simple. Ground effect was a known concept in aerodynamics. It just needed a technical solution to be fully exploited.

Aerodynamic Drag

Drag force is a very large topic in aerodynamics. From a physical point of view, drag is the resultant of forces acting normally and tangentially to a surface, the former ones being pressure terms, and the latter ones viscous terms. The mechanism under which these forces are created is ultimately related to the formation of vortices and shear layers. A car has to be able to overcome drag (the resistance experienced as it travels forwards) and it attempts to do this by presenting the smallest frontal area possible.

By definition drag is force acting on an object in motion through a fluid in an opposite direction to the objects (chassis) motion, produced by friction.

Aerofoil Wings

The simplest inventions are often the best, and never has the role of aerodynamics proved as dramatic as when aerofoil wings burst onto the Grand Prix scene in the late 1960s. These were added in an attempt to provide downforce that gave cars superior traction, and thus made them less likely to spin when being driven around a corner, enabling them to turn at greater speed. Although these aerofoils would alter in shape over the years that followed, they were here to stay.

However, engineers are always faced with a trade-off between maximum downforce and minimum drag, and it's something that varies from circuit to circuit. The need for downforce is at its greatest at circuits with many corners, such as Monaco and the Hungaroring, and the need is least at those with the longest straights such as Monza. Here engineers would angle the aerofoils back to expose the least frontal area in order to reduce drag. This would limit their impact through the corners, but the corresponding aerodynamic efficiency along the circuit's straights would more than make up for this. So, it wasn't surprising that suspension had to be toughened up after a number of breakages. Indeed, these high mounted aerofoils were banned after further breakages and resultant accidents in the 1969 Spanish GP, with aerofoils henceforth having to be lower and sturdier.

Formula 1 Engines

Grand Prix engines are powerful. Everybody knows that. But Grand Prix cars weigh very little too and this combination is what makes the performance of these cars a league away from the performance of even the most high-powered road-going sports car.

A Grand Prix car weighs 600kg including driver. Maximum speeds of 220mph. For example, the fact that the engines in contemporary Grand Prix cars rev to 17,500rpm. This is more than double the amount that most road cars manage before their drivers have to change up a gear or rev their engines past its red line. Their power delivery is also such that peak power comes in at the top end of the rev range, say from 16,000rpm upwards, with relatively power below that. In order to reach this incredible number of revolutions per minute, the favoured cylinder format is one with a large bore and a small stroke. To do it the other way around, that is to say with a small bore and a large stroke would force the engine to work extremely hard to reach those revs. Unsurprisingly, all these revs mean that the engine in the Grand Prix car drinks fuel at an alarming rate. In fact, a Grand Prix car consumes fuel between 180 and 200 litres of fuel per Grand Prix, which translates to close on 1 litre/km, making it five times thirstier for flat-out performance than a family saloon.

While the quest for horsepower is everything, engine builders are also under pressure to make the engines derivable by providing sufficient torque to allow some power lower down the rev range. Having as wide a useable rev range as possible is the aim, especially for use out of tight corners where it's traditionally hard to keep the revs up.

Racing Physics

The Bernoulli effect pays a large role in the operation of the aerodynamic surfaces of an F1 car. The Bernoulli effect is expressed by an equation known as Bernoulli equation. This equation states that, for a given volume of fluid, the total energy is constant. This is due to the principle of the conservation of energy, which applies to most things in racing. When a fluid is in relative motion, this energy is divided up to three parts, the sum of which remains constant for constant external conditions.

The three parts that the total energy can be divided up into are:

1. The pressure energy within the fluid.
2. The kinetic (movement) energy of the air/
3. The potential energy of air (in this case due to elevation).

This can be written as:

p + dpv2 + gh = a constant

Where: p = pressure

d = density of fluid

v = velocity of fluid

g = acceleration due to gravity

and h = height of fluid above some reference point

Since our average racing car does not make a great enough elevation, it changes to make the potential energy of air a variable. We can assume it as a constant and eliminate it from our equation:

p + d v2 = some other constant

This can be written as:

p + q = H

Where: p = static pressure

q = d v2 = dynamic pressure

and H = some other constant

What this all boils down to start of the dynamic pressure increase, then the static pressure must have a corresponding decrease and vice versa. This means that: if a fluid (gas or liquid) flows around an object at different speeds, the slower moving fluid will exert more pressure than the faster moving fluid on the object. The object will then be forced toward the faster moving fluid. The shape of the chassis is similar to an upside down airfoil. The air moving under the car moves faster than that above it, creating downforce or negative lift on the car. Downforce is necessary in maintaining high speeds through the corners by forcing the car to the track. Light planes can take off at slower speeds than a ground effect car can generate on the track. In addition, the shape of the underbody (an inverting wing) creates an area of low pressure between the bottom of the car and the racing surface. This sucks the car to the road, which results in higher corner speeds.

The total aerodynamic package of the car is emphasised now more than ever before. Teams that plan on staying competitive use track testing and wind tunnels to develop the most efficient aerodynamic design. The focus of their efforts is on the aerodynamic forces of negative lift or downforce and drag. The relationship between drag and downforce is very important. Aerodynamic improvements in wings are directed at generating downforce on the race car with a minimum of drag. Unwanted drag, which accompanies downforce, will slow the car. The efficient design of a chassis is based on a downforce/drag compromise.

The necessary downforce is concentrated in three specific area of the car. The ongoing challenge for team engineers is to fine-tune the airflow around these areas:

1. Front wing assembly
2. Chassis
3. Rear wing assembly

Front Wing Assembly

The front wing assembly is constructed of carbon fibre and is the first part of the car to meet the air mass. The fluid field here is better than other parts of the car because the air here has been distributed at the least. The wing is designed to produce downforce and guide the air past the wheels to the radiator inlets and the underbody. The turbulent flow moving towards the rear of the car will impact the efficiency of the rear wing. The efficiency of the wing is based on the following:

The Aspect Ratio

The amount of downforce produced by a wing is being determined by each size. The larger the wing the greater the downforce. The length/width ratio is called the aspect ratio. The higher the aspect ratio the more efficient is the wing. The higher the aspect ratio the less air resistance created by the vortex at the wing tips. The aspect ratio is the span of the airfoil (the long dimension perpendicular to the airflow) divided by its chord (the dimension parallel to the airflow).

The Angle of Attack

The efficiency of a wing is its downforce/drag ratio. The amount of downforce generated depends on the angle or the tilt of the wing. The greater the angle of attack the more downforce is being generated.

Drag

The drag increases with the angel of attack. The downforce generated by the wing works in a vertical downward direction, while the drag acts in the opposite direction to the airflow.

In setting up the front wing assembly, engineers must consider what happens to the airflow as it travels toward the back of the car. In an effort to clean up distributed airflow, small adjustments are made to the front of the car. Turning vanes are used to deflect the turbulent flow wake away from the inside of the front wheels and front wing. Small vertical wing sections mounted on the front wing assembly are used only on the street and road set-ups. This will not be efficient on the speedway configuration due to the increase in drag at higher speeds. The front wings on the speedway set-up are very small and function as trimming devices rather than creating downforce. The front wing assembly is capable of producing 1500 Ibs of downforce.

The front and the rear wheels create the greatest source of drag of the F1 car. This can account up to 60% of the aerodynamic efficiency of the car. The front wing assembly has end plates attached at 90 degrees to the end of the wings. The end plates help in reducing turbulence around the front wheels and assist airflow moving to the radiators and around the sidepods.

Any change made to the front of the car will affect the airflow moving toward the rear of the car. The team engineers must consider how any change to the front of the car will impact the overall aerodynamic efficiency of the car. The aerodynamic set-up of the car has been refined to the point that minute changes can improve the "ground effects" of the car. Air directed under the chassis creates an area of low pressure between the underbody of the car and the racing surface. The car is basically sucked down to the road, which results in higher corner speeds.

Chassis

The F1 chassis is designed to produce maximum downforce, while at the same time it minimises the drag. The top of the car is designed to slice through the air, while the underbody is shaped to create an area of low pressure between the underbody and the track. The shape of the tunnels under the car resembles an inverted wing. The shape of the underbody and the use of the front and rear airfoils create the "ground effect" of the car.

Safety regulations restrict the size and height of these wings and soon designers were seeking another source of downforce. This led to the redesign of the underbody of the car. The sidepods of the car housing the radiators were designed to allow airflow under the car through the tunnels. The tunnels are narrowed in the centre and expanded toward the rear of the car. Because of this an area of low pressure was created between the underbody of the car and the track. This resulted in the car being forced down to the track. F1 designers included skirts on the cars to reduce lateral airflow under the car. Skirts were attached to the sidepods and were vertically adjustable to seal off lateral flow.

The underbody is of major importance in the aerodynamic set-up of the car. The frontal area of the car is reduced in size so it can slice through the area, in an effort to reduce the drag coefficient. The front wings are not used for downforce on the speedway set-up, but rather as a trimming device. The speedway rear wing is shaped to create less downforce and reduced drag. Drag reduction is more important than downforce on the speedway set-up. The best set-up would be an adjustable wing to reduce drag on the straights but supply downforce in the corners.

The Rear Wing

The rear wing is made of major carbon fibre and is attached to the transmission housing. Low downforce, standard and high downforce wings are the choices under current regulations. The objective is to achieve the best downforce/ drag compromise. It provides maximum downforce but also a maximum amount of drag.

As the airflow moves towards the rear of the car it becomes more turbulent. The wake from the front wings, mirrors, the driver's helmet, the front wheels and the sidepods all influence airflow. All of these obstructions create a turbulent airflow towards the rear of the car. The rear wing in not as aerodynamically efficient as the front wing yet it must generate more than twice as much downforce to balance the car, thus, the rear wing assembly is designed to produce high downforce. Because of the configuration of the wings, unwanted drag is also creating at high speeds. The most efficient aerodynamic set-up of the race cars is based on a downforce/drag compromise.

Engine-Ram Air Intakes

F1 engineers are constantly at work in order to improve the engine use air intakes performance. Based on "a good supply of air to the engine air trumpets, they can lead to significant improvement in engine performance".

Ram air intakes occur when a vehicle is in motion. The pressure within the air intake can increase by an amount equal to the ram air pressure/dynamic of the undisturbed stream.

Dynamic pressure = d v2

= air density

v = air velocity

To obtain true ram effect, the air has to be taken cleanly from the free stream without friction or turbulence losses. Recent designs have shaped the air intake to reject the turbulent boundary layer from ahead surfaces; for instance the engine intake above the drivers head on an F1 car reject turbulent flow from driver's helmet (as shown in the following picture). It is best to direct the air to a large volume where air speeds are low. This ensures that the airflow evenly distributed to all engine funnels. This is called an airbox.

The NACA Duct

When a side rather than a forward facing is required for the induction of air, the NACA duct can be used (as seen in the figures below). The theory of the NACA duct is that, when it takes air from the free stream without energy losses, it is just efficient as a forward facing ram effect aperture. It is deal to place a NACA duct near the front of a surface, where the boundary layer is thin and laminar. The edges of the vent must be sharp; this allows the airflow to flow up around these edges to form a vortex along each side. The pressure differentials of the vortices help draw air in the vent. NACA duct is more frequently used for small volume cooling flows such as break or oil cooling.

CHAPTER 5
CORROSION IN HEAT EXCHANGERS

Heat exchangers are commonly used to transfer heat from steam, water, or gases, to gases, or liquids. Some of the criteria for selecting materials used for heat exchangers are corrosion resistance, strength, heat conduction, and cost. Corrosion resistance is frequently a difficult criterion to meet. Damage to heat exchangers is frequently difficult to avoid.

There are many variables related to specifying heat exchangers —for example, temperatures, flow rates, pressure drop and heat transfer—that are common to all types. There also are some variables that are specific to specifying a plate heat exchanger, in particular, the number of passes.

An example of a heat exchanger is shown below:

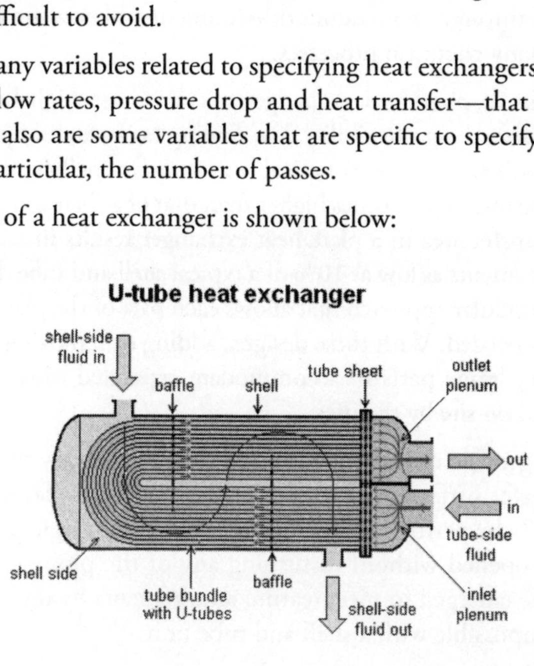

The main components of the plate heat exchanger include:

- Plate Pack. The plate pack is composed of a number of corrugated heat transfer plates according to the heat transfer surface required. The plate corrugations provide support between plates and encourage turbulence, which gives the plate heat exchanger its high heat transfer.
- Gaskets. The molded elastomeric gaskets on the plates ensure that the flow channels are sealed securely from each other, preventing cross-contamination. The gaskets also determine the direction of flow within the exchanger.
- Frame. The frame provides structural support and pressure containment by enclosing the plate pack. The main frame components consist of a fixed end, a movable end, upper and lower carrying bars, and tightening bolts.
- Connections. The connections for incoming and outgoing media usually are found in the fixed end plate. In the case of a multiple-pass design, connections also are found on the movable end.

The plate heat exchanger is well known as one of the most efficient heat exchanger designs. Its basic concept, using corrugated plates for heat transfer, was introduced in the early 1930s for the food and dairy industry. Since then, plates have gone through tremendous development and now are used as standard equipment in many chemical processes.

The hallmarks of plate heat exchanger operation are high heat transfer efficiency, compact size, design flexibility, low fouling (as a result of plate pattern turbulence) and low lifecycle cost. A typical heat transfer coefficient for a plate heat exchanger may be four to five times higher than that of a shell and tube design. The compact heat transfer area in a plate heat exchanger results in an efficient design with space requirements as low as 10% of a typical shell and tube design. Some designs employ a modular approach that allows each part of the plate heat exchanger to be changed as needed. With these designs, adding or removing plates for a new duty or changing frame parts to accommodate expanded capacity requirements can be performed on site by the user.

Plate exchangers are quickly opened and closed. It is easy to get at and clean all the heat transfer surfaces and to replace any damaged plates and/or gaskets. This can normally be performed by one individual. On single pass units, the exchanger can be opened without disturbing any of the piping. In addition, the exchanger can be enlarged to meet feature requirements by the addition of extra plates. This is impossible with a shell and tube unit.

A Plate Heat Exchanger can work with much closer temperature approaches, and the same heat load with a much lower flow rate, thereby saving pumping or operational costs. It is possible to cool a liquid down within 2°F of the hot medium. Heat recovery of up to 85% to 90% is quite common, making the plate heat exchanger the most efficient heat exchanger for heat recovery, particularly for applications where the temperature difference between the media is small.

Industries and applications where heat exchangers are used:

- Automotive
- Brewing
- Caustic Soda
- Chemical
- Textile
- Food
- HVAC
- Marine
- Metal Working
- Petroleum
- Power
- Pulp and Paper
- Steel

The plate heat exchanger has ports for the passage of two fluids (as shown above) between which the heat transfer will occur (as shown above). The plate pack is assembled between two pressure-retaining frames compressed with tightening bolts.

Corrosion Problems/Prevention/Advantages

Heat exchangers are designed to contain liquids, gases and solids such that a loss of containment does not occur. Other examples are shown on the following page.

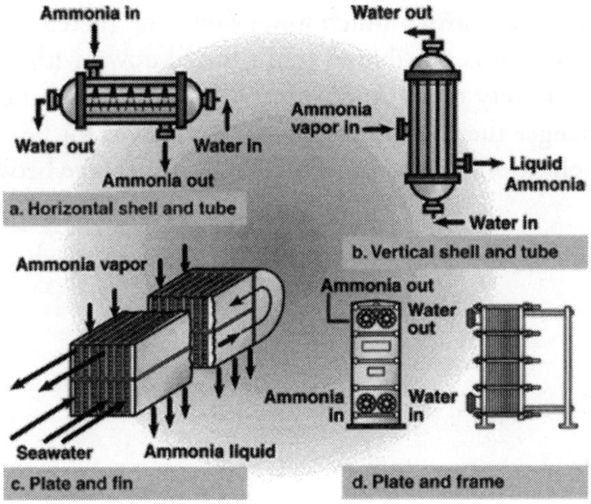

Leaks or the mechanical or structural failure of these items of equipment may result in a major accident on-site. The presence of flaws in the critical components which are introduced during the fabrication processes (for example lack of fusion), during construction, commissioning or the effect of degradation during normal operation (for example corrosion, chemical degradation and fatigue) may result in the integrity of such systems being compromised and increase the likelihood of failure.

Heat exchangers, unlike mechanical devices, rarely fail, but if misused that can suffer from failure like any other product. There is seldom any visible warning of possible problems, unlike a machine which runs incorrectly and shows obvious signs. Generally speaking, the majority of problems facing a heat exchanger are those of corrosion or eroding on the water side. There are basically four types of problems. These are:

1. Impingement attack (or Corrosion/Erosion)

This is caused by water containing air bubbles flowing at a high speed through the tubes. The impingement of rapidly moving water may lead to a breakdown of the protective film, so allowing corrosion/erosion of the tubes. The effect of these conditions would be pockmarking and pinholing of the tube.

The pressure loss of a flowing fluid due to friction varies approximately with the square of the flow velocity. As the velocity increases, the abrasive effect on the tube wall increases, and erosion of the tube may occur. The extent of erosion caused by excessive velocity is dependent upon the physical characteristics of the tube material or impediments to flow on the tube wall, such as burrs on the tube ends, blobs of solder, or mineral deposits.

Erosion corrosion occurs at locations where turbulence develops in a system. This turbulence interferes with the normal protective film formation on the inside of the tube and also erodes the copper surface at that point. Turbulence can be caused by excessive velocity, sudden changes in direction and flow obstacles such as burrs and solder blobs.

Erosion corrosion is readily identified from the characteristic appearance of the damaged surface. A typical damaged surface has deep horseshoe pits with the open ends facing downstream.

2. Oxide Corrosion

This is caused by water containing organic matter such as that found in polluted estuaries, usually this water produces hydrogen sulfide, which is very corrosive and can cause failure of the tubes, particularly if excessive water flows are used.

Hydrogen sulfide is a toxic, colorless gas that is odorless at high concentrations but smells like rotten eggs in low concentrations. Hydrogen sulfide is produced during the decomposition of organic matter and occurs with hydrocarbons in some areas. Hydrogen sulfide is a serious and potentially lethal hazard in several regions of hydrocarbon exploration and production, so awareness, detection and monitoring of hydrogen sulfide is essential. It is also corrosive, requiring costly special production equipment such as stainless steel tubing.

3. Pitting

This problem is caused by very aggressive sea water in the tubes, especially in partially filled coolers where the sea water is stagnant. Low sea water flow rates can create a high temperature rise on the sea water side this allows salts to come out of solution and contributes to the corrosion problems. Under these conditions deposits may build or settle in the tube, allowing pitting-corrosion, to take place under the deposit.

Pitting is a localized form of corrosive attack. Pitting corrosion is typified by the formation of holes or pits on the metal surface. Pitting can cause failure due to perforation while the total corrosion, as measured by weight loss, might be rather minimal. The rate of penetration may be 10 to 100 times that by general corrosion.

Pits may be rather small and difficult to detect. In some cases pits may be masked due to general corrosion. Pitting may take some time to initiate and develop to an easily viewable size.

Pitting occurs more readily in a stagnant environment. The aggressiveness of the corrodent will affect the rate of pitting. Some methods for reducing the effects of pitting corrosion are listed below:

- Reduce the aggressiveness of the environment
- Use more pitting resistant materials
- Improve the design of the system

4. Crevice Corrosion

Crevice corrosion is a localized form of corrosive attack. Crevice corrosion is occurs at narrow openings or spaces between two metal surfaces or between metals and nonmetal surfaces. A concentration cell forms with the crevice being depleted of oxygen. Some examples of crevices are listed below:

- Flanges
- Deposits
- Washers
- O-rings
- Gaskets
- Sediment

Some methods for reducing the effects of crevice corrosion are listed below:

- Eliminate the crevice from the design
- Select materials more resistant to crevice corrosion
- Reduce the aggressiveness of the environment

Fault Finding

There are four conditions which can give an indication of possible problems with heat exchangers. These have been listed below against the causes and the remedy required:

Symptoms

- Increase in temperature on the shell side or excessive pressure loss.
- The pressure loss is as expected, but the temperature of prices Process fluid leaking into the cooling water circuit, or cooling water leaking into the process circuit.
- Process flow rate or cooling water circulation appears to be incorrect.

Possible Causes

- Oil sludging, tube scaling or a build up of both resulting in a insulating film covering the tubes.
- Film, scale or restrictions on the inside of the tubes.
- Split or perforated tubes
- General lack of process fluid or water in the system. Process fluid or water pump performing incorrectly.

Remedy

- The complete heat exchanger should be thoroughly cleaned.
- Again, remove heat exchanger and clean thoroughly.
- As a temporary measure the tubes affected should be plugged or closed with hard wood plugs. The tube stack should be replaced as soon as possible by a new unit, together with new 'O' ring seals.
- Add process fluid or water to make up the correct levels. Examine each unit and rectify as necessary.

Advantages

There are several advantages to using plate heat exchangers, including:

- Reduce Fouling. Fouling is reduced due to high turbulence, uniform liquid flow, and smooth plate surfaces. The elimination of dead spots and the lack of corrosion products both contribute to low fouling. This enables the plate exchanger to achieve higher water outlet temperatures than the equivalent shell and tube without the need to clean frequently.

- Cost Savings. Plate heat exchangers save costs with their high efficiency, low first cost, compact installation and ease of maintenance.
- High Heat Transfer Coefficients. Plate heat exchangers provide high rates of heat transfer due to the turbulence-inducing corrugations of the plates. Different corrugations allow the plate heat exchanger to be tailored to specific heat transfer and pressure drop requirements. With water vs. water type applications, it is not unusual to achieve 4oF approach temperatures.
- Safety. The plates' special gasket configuration prevents any mixing of the media being processed. In the port area, the fluids are separated by a double-gasket design with a vent to atmosphere, where the fluid will be released if one of the gaskets fail. In more extreme cases, a safety plate design can be provided. By welding two plates together, a safety barrier is provided so that one fluid drains to atmosphere instead of contaminating the adjacent fluid if a plate fails.
- Flexibility. Plate heat exchangers can be adapted to changing process conditions by simply adding or removing plates. This eliminates the need for reinvestment.
- Low Product Hold Volume. The small volume of liquid in the plate heat exchanger is the reason for the relative low weight of the equipment. It enables a faster startup and shutdown of the complete plant compared to conventional heat exchangers. Lower hold volumes also mean higher recovery rates for high value products.
- Maintenance. Any maintenance of the plate heat exchanger can be carried out easily. Where a shell-and-tube heat exchanger requires a clear pull space equal to the length the tube bundle, a plate heat exchanger only needs a few feet clear on each side of the unit for all maintenance.
- Materials. Plates are available in several grades of stainless steel, including 304, 316, 317, and in specialty materials such as Hastelloy, Incoloy, titanium, nickel and tantalum. The gaskets are available in nitrite rubber (NBR), butyl rubber (Butyl), ethylene propylene rubber (EPDM), silicon rubber and Viton.
- Low Risk of Leaks: The double gasket sealing system, laser wielded twin-plates, and corrosion-resistant materials, prevent inter-leakage between the media, making it virtually impossible for ammonia to get to the brine. In the rare event of even the slightest leak, it would be external and easy to detect. The small refrigerant volume allows for rapid emptying and opening the unit. The faulty plate or gasket could easily be removed or replaced, and

the plate pack quickly reassembled. Leakage in shell-and-tube chillers could go unnoticed for a long period of time, causing severe damage to the system, and take days for expensive repairs to be completed.

- Rolled threads and zinc plating the bolts gives a stronger, corrosion resistant thread.

CHAPTER 6
NON-DESTRUCTIVE TESTING FOR WELDING

Non-destructive testing (NDT) for welding is a term used for the examination of welded joints in a way that allows them to be examined without changing or destroying them.

NDT plays an important role in everyday life and is necessary to assure safety and reliability. Examples are found in aircraft, motor vehicles, pipelines, bridges, trains, power stations; refineries and oil platforms which are all inspected using NDT.

NDT is a Quality Assurance management tool which can give impressive results when used correctly. It requires an understanding of the various methods available, their capabilities and limitations, knowledge of the standards and specifications for performing the tests.

Welded joints which fail to achieve their design requirements or projected life because of undetected defects may require expensive repair or early replacement, Such defects may also be the cause of unsafe conditions or terrible failure, as well as loss of profits because of unplanned plant shutdown.

Non-destructive testing can be applied to each stage of an items construction. The materials and welds can be examined using NDT and either accepted, rejected or repaired. NDT techniques can then be used to monitor the integrity of the item or structure throughout its design life.

Testing Methods

The method that can be used for non-destructive testing depends on the physical properties of the material. A thorough knowledge of each NDT method is required to ensure the correct selection of the appropriate method for each application.

NDT is used for the following reasons:

1. Accident prevention.
2. To reduce costs.
3. To improve product reliability.
4. To determine acceptance to a given requirement.
5. To give information on repair criteria.

NDT can be used to find, size and locate surface as well as subsurface defects.

The NDT methods studied in this chapter are:

- Visual Examination.
- Liquid Penetrant.
- Magnetic particle.
- Radiography.
- Ultrasonic.

Visual Inspection

Visual inspection is the most common type of inspection. It is inexpensive and normally doesn't require special equipment. This method detects defects and discovers possibly fabrication problems. Visual inspection should be performed before, during and after welding operations to detect defects that can only be found later by more expensive NDT methods.

Equipment

Visual aids and gages are sometime used to make detection of defects easier and to measure the size of the defects in the welds. If the area to be inspected is not readily visible, the inspector may use mirrors, borescopes, flashlights or other aids.

Inspection of weld joints includes quantitative as well as qualitative assessment of the joint. Standard measuring tools are available to make various measurements, such as joint geometry and fit-up, weld size, weld reinforcement height, misalignment and depth of undercut.

Before Welding

Examination of the base metal before fabrication can detect conditions that can cause weld defects. Scabs, seams, scale can be detected with visual inspection. Plate lamination may be observed on cut edges. Dimensions should be confirmed by measurements. Base metal should be identified type and grade. Corrections should be made before work proceeds.

After the parts are assembled for welding, the weld joint root opening, edge prep and other features should be checked that might affect the quality of the weld. The inspector should check conditions for conformity to applicable specifications.

List of check points

1. Joint prep, dimensions and cleanliness
2. Clearance dimensions of backing strips, rings or consumable inserts
3. Alignment and fit-up
4. Welding process and consumables
5. Welding procedures and machine settings
6. Specific preheat temp
7. Tack weld quality

During Welding

During welding visual inspection is the main method of quality control.

List of check points

1. Treatment of tack welds
2. Quality of root pass and succeeding weld layers
3. Proper preheat and interpass temp
4. Sequence of weld passes

5. Interpass cleaning
6. Root condition prior to welding second side
7. Distortion
8. Conformance with applicable procedure

The most important part of a weld is the root pass. Complete visual inspection of the root pass may detect conditions that can cause a defect in the completed weld. The root opening should be monitored as the welding progresses. The tack welds, clamps or braces should be designed to maintain specific root opens to assure proper joint penetration and alignment.

Inspection of successive layers of weld metal is concentrated on the bead shape and cleaning. When specified, preheat and interpass temperatures should be monitored at the proper times with a suitable temp measuring device. The amount of heat input and also the sequence and placement of each weld pass may be specified to maintain mechanical properties or limit distortion or both. Each weld pass should be evaluated before the next bead is applied.

After Welding

Items that are checked after welding are:

1. Final weld appearance
2. Final weld size
3. Extent of welding
4. Dimensional accuracy
5. Amount of distortion
6. Postweld heat treatment

Most codes and specifications describe the type and size of discontinuities that can be accepted. Many of the following defects on the surface can be found by visual inspection:

1. Cracks
2. Undercut
3. Overlap

4. Exposed porosity and slag inclusions
5. Unacceptable weld profile
6. Roughness of the weld

For detection and accurate evaluation of defects, the weld surface should be thoroughly cleaned of oxide and slag, which if not can cause the weld to be rejected by most codes. Dimensional accuracy of the weld is determined by conventional methods. The conformity of the weld size and contour may be detected by using suitable weld gages. The size of a fillet weld, whose parts are at right angles, or near, is defined in terms of the length of legs. The gage should determine if the leg size is within the allowable limits, and whether there is excessive concavity or convexity.

For bevelled welds, the height of reinforcement should be consistent with requirements. Surface appearance requirements differ widely. The weld surface appearance should meet the requirements of the standard. A fabrication standard may permit limited amounts of undercut, undersize and piping porosity, but cracks, incomplete fusion and unfilled craters are generally not accepted. Undercut, overlap and improper weld profile act as stress risers under load, and cracks may develop at these locations under cyclic loads.

When a postweld treatment is specified, the operation should be monitored. Items of importance are:

1. Area to be heated
2. Heating and cooling rates
3. Holding temp and time
4. Temp measurement and distortion
5. Equipment calibration

The main disadvantage of visual inspection is that it can't detect subsurface defects, but proper visual inspection procedures before and during fabrication can increase product reliability.

Liquid Penetrant Testing (PT)

Liquid Penetrant testing is a method that exposes open defects to the surface by bleedout of a liquid penetrant medium against a contrast background developer. The technique is based on the ability of a penetrating liquid to wet the surface opening of a defect and to be drawn into it. If the defect is major, penetrant will be held in the cavity when the excess is removed from the surface. Upon application of a liquid propeller or dry powder developer, blotter action draws the penetrant from the defect to provide a contrasting indication on the surface.

The following steps are normally used in application of a typical penetrant test:

1. Clean the test surface
2. Apply the penetrant
3. Wait for the prescribed dwell time
4. Remove excess penetrant
5. Apply the developer
6. Examine the surface for indications and record the results
7. Clean to remove the residue

Liquid penetrant methods are divided into two groups:

1. Fluorescent: The penetrant medium is fluorescent, meaning it glows when illuminated by ultraviolet or "black" light
2. Visible: Uses a visible penetrant, usually red in colour that produces a contrasting indication against a white developer background.

Magnetic Particle Testing

Magnetic particle testing is used to detect surface or near surface defects in magnetic materials. The method is based on the principle that magnetic lines of force in a ferromagnetic material will be distorted by a distinct change in material continuity. A defect or sharp dimensional changes are examples of such a change. If a defect in a magnetized material is open to or close to the surface, the magnetic flux lines will be distorted at the surface, called flux leakage. When fine magnetic particles are distributed over the area of the defect while the flux leakage exists, they will build up at the defect area and be held in place.

The essential requirements of the test method are:

1. The part must be magnetized
2. Magnetic particles must be applied while the part is magnetized
3. Any accumulation of magnetic particles must be observed and interpreted

A ferromagnetic material can be magnetized either by passing an electric current through the material or by placing the material within a magnetic field originated by an external source. The entire component, or a part of it, can be magnetized as dictated by size and equipment capacity by need. If a defect is open to the surface, the flux leakage will be a maximum for a given size and shape of defected. When a defect is below the surface, flux leakage will be lower. Defects must be open to the surface or must be in the near subsurface to create flux leakage of sufficient strength to accumulate magnetic particles.

If a defect is formed nearly parallel to the lines of force, it will be undetectable. Because defects may occur in any place on the welded joint, it is usually necessary to magnetize the part at least twice so that induced magnetic lines of force are produced in different directions to perform an adequate examination.

The difference in sensitivity is mainly because of the fact that the can discern the contrast of a fluorescent indication under black light more readily than a colour contrast under white light.

Ultrasonic Inspection

Ultrasonic testing uses waves above the frequency limit of the human being hearing ability and usually in the frequency range 0.6 to 5 MHz. A pulse consisting of a number of these waves is projected into the specimen under test. If a defect exists in the tested joint an echo is reflected from it and from the type of echo the kind of defect that exists can be figured out.

The equipment includes an electrical unit which generates the electrical oscillations, a cathode ray tube where the pulse and echo can be seen, and probes which introduce the waves into the specimen and receive the echo. The electrical oscillations are converted into ultrasonic waves in a transducer which consists of a piezo-electric element fitted in a perspex block to form the probe, which, when in use, has its one face pressed against the surface of the material under test. When a pulse is injected into the tested joint a signal is made on the cathode ray tube. The echo from a flaw is received by another probe, converted to an electrical e.m.f. (varies from microvolts to several volts) by the transducer and is applied to the cathode ray tube where it can be seen as a signal displaced along the time axis of the tube from the original pulse.

When the ultrasonics was first used to detect defects in welds, it used to employ longitudinal waves projected into the specimen at right angles to the welded joint. This caused problems because it meant that the weld surface had to be dressed smooth before examination, and not the way which the defect oriented, The type of wave used to overcome these disadvantages is introduced into the specimen with a bit of distance from the welded joint and at an angle to the surface could be 20° for example, and is known as a shear wave. The frequency of the waves (are usually 2.5 and 1.5 MHz for butt welds), the angle of incidence of the beam, the type of surface and the grain size, all affect the intensity of the echo which is adjustable by a sensitivity control.

There are three types of probes:

1. A single probe which acts as transmitter and receiver, the same piezo-electric elements transmitting the pulse and receiving the echo. The design of the probe is complicated in order to prevent reflections within the perspex block confusing the echo.

2. The twin transmitter-receiver probe in which transmitter and receiver are mounted together either side by side or one in front of the other but are quite separate electrically and ultrasonically so that there is no trouble with interference with the echo. This type is the most popular.

3. The separate transmitter and receiver each used independently (two-handed operation).

To make a 'length scan' of the weld the transmitter-receiver unit is moved continuously along a line parallel to the welded seam so that all points of the whole area of the welded joint are covered by the scanning beam and care must be exercised that by the use of too high a spread of the beam, double echoes are not obtained from a single defect. It is clear that changing the distance from the weld to the probe changes the depth at which the main axis of the beam crosses the welded joint and moving the probe at right angles to the line is known as depth scan. A spherical defect will have no directional characteristics and a wave falling upon its centre will be reflected along the incident path, the amplitude of the echo depending upon the size of the defect. Cylindrical flaws behave in the same way but in the case of a narrow planar defect it is evident that optimum echo will be received when the crack is at right angles to the wave and there will be no echo when the crack lies along the wave, but if the probe is moved to the first echo position the crack is no longer lying along the beam.

The probes must make good contact with the specimen and on slightly curved surfaces a thin film of oil is used to improve the contact. On surfaces with bigger curves, like when testing welds on pipes, curved probes are used.

X-ray Testing

X-rays are created in the following way: A glass tube with two arms and a central bulb is exhausted of air. In one arm is a small electric light filament (F) which can be heated to white heat. In the other arm is a thick copper stem (S) ending in a target (T) made of platinum and inclined at an angle of 45o to the axis of the tube. A high voltage of between 60000 and 180000 volts is placed across the ends of the tube; the end (A) being positive (called the anode). The fact that the filament is white hot, and emits negatively charged particles, enables the high voltage to send a current through the tube, and the current causes a stream of negatively charged particles, called cathode rays, to move from the filament (F) to the positively charged target (T), as shown. On hitting the target, they are reflected as shown and are then termed X-rays.

They are electro-magnetic radiations of short wave-length and they can penetrate solid materials, but a just a few of rays is absorbed in the process. The amount of absorption depends on the thickness of the material and on its density. The denser and thicker the material, the smaller the proportion of rays which get through. Certain materials, such as calcium tungstate and barium platinocyanide, become fluorescent or luminous when X-rays strike them. If a screen is coated with one of these materials and the rays fall on it, the sensitized screen becomes brightly illuminated. If now an object, such as a piece of steel, is placed in the path of the rays in front of the screen, a shadow of the object will be thrown on the screen, and if the object is of the same thickness and density throughout, the shadow will also be of the same degree of darkness over all its surface. If, however, any holes or cavities exist in the steel, the rays will be less absorbed at these points and the shadow will be more brightly illuminated at these spots.

In the same way, films covered with silver halides are affected by X-rays in the same way that ordinary photographic films are affected by light. Thus if an X-ray film is put in place of the sensitive screen in the above, a `shadowgraph' of the object will appear when the film is developed. Films are used more than the screen method, because they provide a permanent record of the shadow which can be carefully studied.

Like photography an incorrectly exposed film will possess no detail, so with an X-ray photograph no detail of defects in the object can be observed unless correct exposure is given. This is entirely a matter of practice.

In order to make sure that we are getting a correctly exposed negative, so that even the smallest defects will be shown up, it is usual to place a penetrometer, which is a small strip of cutting of the same material as the object for example steel and about 1/100 to 2/100 of its thickness, on the upper surface next to the X-ray tube.

If this appears as a shadow on the negative, we are sure that any defects or holes of the size 1/100 to 2/100 of the thickness of the object will be indicated.

Great practice is necessary to interpret the X-ray films of welds correctly and to distinguish between various defects shown up as shadows. Gas holes causing porosity arc usually regular in shape, while any included slag is usually very irregular. In this way, we can determine whether penetration to the full depth required has been obtained, whether correct fusion between parent metal and weld metal or between layer and layer in a multi-layer weld has been obtained, and whether there are regions of entrapped slag, blowholes or other porous defects. In addition, any defects, such as contraction cracks, will also be shown up clearly.

The X-raying of butt welds is comparatively straightforward, but with fillet welds special methods have to be adopted, and often more than one photograph taken in different directions. Indicates some of the defects mentioned above and how they appear in the X-ray photograph.

Finally, the method of X-ray inspection of pressure vessels, such as boilers, may be mentioned. During the welding operation on the boiler, one or more test pieces are made by the welder, using the same technique exactly as on the boiler, the test pieces being of the same metal and thickness as the boiler itself. The weld on the boiler is X-rayed along its whole length and the test pieces are also X-rayed. A comparison of the negatives will indicate how closely the structure of the test pieces compares with that of the boiler itself and if this is very close, it can reasonably be supposed that the boiler weld would behave as the test pieces under test. The test pieces are then tested by the usual methods and finally tested to destruction. This affords an excellent indication as to the probable behavior of the boiler.

CHAPTER 7

MANUAL METAL ARC WELDING (MMA) & GAS METAL ARC WELDING (GMAW)

Manual metal arc (mma) welding was first invented in Russia in 1888. It involved a bare metal rod with no flux coating to give a protective gas shield. The development of coated electrodes did not occur until the early 1900s when the Kjellberg process was invented in Sweden and the Quasi-arc method was invented in the UK. But the coated electrodes used these days were slow to be adopted because of their high cost. But because the need for strong welds grew, manual metal arc and coated electrodes became a unit; Stainless steel, inconel, nickel and cast iron electrodes are also commonly used, as are surfacing type electrodes for the building up of wear resistant surfaces.

Mma is widely used in the fabricating industries for the construction and repair of plain carbon and low alloy steels. The equipment is of initial low cost, and the process is flexible and safe to use. Good ventilation and protective clothing is essential.

Manual metal arc welding is known by many names depending on the country and the region; some of them are listed below:

- MMAManual
- STICKManual

- ELECTRIC ARCManual
- SMAWShielded Metal Arc Welding

Description of the Process

Mma consists of an AC/DC arc burning between a consumable electrode, when an arc is struck between the metal rod (electrode) and the work piece, both the rod and work piece surface melt to form a weld pool, Instant melting of the flux coating on the rod will form gas and slag which protects the weld pool from the surrounding atmosphere. The slag will solidify and cool and must be chipped off the weld bead once the weld run is complete (or before the next weld pass is deposited).

Equipment

The basic equipment consists of:

- Power source: Requires a controllable supply of AC/DC current at between 50 and 500 amps, and an arc voltage of 20 to 50 volts. Open circuit voltage must be high enough to initiate the arc, (60-100 volts), therefore a drooping or constant current power source would be necessary.
- Electrode holder: Designed to protect the operator from the currents and voltages used during welding and to enable the electrode to be securely fastened in position, to make the access to joints easy. The electrode holder is connected to the power source.
- Cables (Welding lead return): Designed with heavy insulation to carry current and also to protect from electric shock, fixed with a secure metal clamp to connect to the work piece from the power source.

Control of the Process

The control of this welding process depends on:

- The Operator
- The Current

These are the main parameters for the MMA process.

Current range is dependent on the size and type of electrode used; and the thickness of material and welding position. A high level of manual dexterity is required to co-ordinate the electrode, to match the burn off rate and to maintain a constant arc length. The quality of the welded joint depends entirely on the operator's skill.

The characteristics of the process are determined by the following features:

- Arc characteristics
- Shielding
- Weld pool control
- Alloying elements

Arc Characteristics

To improve metal transfer and reduce spatter, certain chemicals are added to the coating. These chemicals also reduce operating voltage for the electrode.

Shielding

This is provided by gases produced by the flux and slag covering during welding.

Weld Pool Control

Slag fluidity determines the ease of positions welding. Fast freezing slags are more suitable for welding in the vertical and overhead positions.

Alloying Elements

The flux coating can contain elements which can improve mechanical properties of the joint, without changing the core wire composition. This is a more economical proposition than changing the core wire.

Types of Flux/Electrodes

Arc stability, depth of penetration, metal deposition rate and positional capability are greatly influenced by the chemical composition of the flux coating on the electrode. Electrodes can be divided into three groups:

- Basic (calcium carbonate and fluoride)
- Rutile (titanium dioxide)
- Cellulosic (cellulose)

Basic

Basic electrodes contain a high proportion of calcium carbonate (limestone) and calcium fluoride (fluorspar) in the coating. This makes their slag coating more fluid than rutile coatings; this is also fast-freezing which assists welding in the vertical and overhead position. These electrodes are used for welding medium and heavy section fabrications where higher weld quality, good mechanical properties and resistance to cracking (due to high restraint) are required.

Features

- low weld metal produces hydrogen
- requires high welding currents/speeds
- poor bead profile (convex and coarse surface profile)
- slag removal difficult

Rutile

Rutile electrodes contain a high proportion of titanium oxide (rutile) in the coating. Titanium oxide promotes easy arc ignition, smooth arc operation and low spatter. These electrodes are general purpose electrodes with good welding properties. They can be used with AC and DC power sources and in all positions. The electrodes are especially suitable for welding fillet joints in the horizontal/vertical (H/V) position.

Features

- moderate weld metal mechanical properties
- good bead profile produced through the viscous slag
- positional welding possible with a fluid slag (containing fluoride)
- easily removable slag

Cellulosic

Cellulosic electrodes contain a high proportion of cellulose in the coating and are characterised by a deeply penetrating arc and a rapid burn-off rate giving high welding speeds. Weld deposit can be coarse and with fluid slag, deslagging can be difficult. These electrodes are easy to use in any position and are noted for their use in the 'stovepipe' welding technique.

Features

- deep penetration in all positions
- suitability for vertical down welding
- reasonably good mechanical properties
- high level of hydrogen generated ; risk of cracking in the heat affected zone (HAZ)

Metal Powder

Metal powder electrodes contain an addition of metal powder to the flux coating to increase the maximum permissible welding current level. Thus, for a given electrode size, the metal deposition rate and efficiency (percentage of the metal deposited) are increased compared with an electrode containing no iron powder in the coating. The slag is normally easily removed. Iron powder electrodes are mainly used in the flat and H/V positions to take advantage of the higher deposition rates. Efficiencies as high as 130 to 140% can be achieved for rutile and basic electrodes without marked deterioration of the arcing characteristics but the arc tends to be less forceful which reduces bead penetration.

Care and Storage of Electrodes

The quality of the welded joint and ease of welding depends entirely on the condition of the electrodes. If the mineral coating is damaged, poor arc stability and inadequate shielding will result.

If not properly protected most of the coating materials can absorb water. This could result in the coating deteriorating and subsequent hydrogen pick-up in the weld bead. Special care must be taken with the storage of controlled hydrogen electrodes. The electrodes should be re-dried according to the manufacturers instructions if they absorb damp.

Increased spatter, striking difficulties weld bead porosity and harsh arcing characteristics can all be indications of poor electrode condition. The condition of the electrode profoundly influences the ease of welding and weld metal quality.

In applications varying from simple fabrications of plain carbon steel to demanding joints in high strength materials, manual metal arc continues to be the process used.

Damage to, or deterioration of the electrode coating may be indicated by any one or more of the following: weld metal porosity; arc instability; excessive spatter; poor bead profile; undercut; poor slag release; heat affected zone cracking.

Therefore for quality and effectiveness attention to the correct storage and handling of electrodes cannot be over-emphasised.

Advantages

1. Good manual application
2. Low process cost
3. Low cost and simple equipment
4. Low cost filler metals
5. Good metallurgical joint properties

Disadvantages

1. Not applicable to all copper based alloys
2. Flux necessary
3. Risk of inclusions and porosity in joint
4. Not capable of a high degree of mechanisation
5. High degree of manual skill required

What's New?

Transistor (inverter) technology is now enabling very small and comparatively low weight power sources to be produced. These power sources are finding increasing use for site welding where they can be readily transported from job to job. As they are electronically controlled, add-on units are available for TIG and MIG welding which increase the flexibility. Electrodes are now available in hermetically sealed containers. These vacuum packs obviate the need for baking the electrodes immediately prior to use. However, if a container has been opened or damaged, it is essential that the electrodes are redried according to the manufacturer's instructions.

Gas Metal Arc Welding (GMAW) or Metal inert gas (MIG) welding was first used in the USA in 1949 for welding aluminium but it didn't become so popular in the UK until about 1952 for welding aluminium using argon as the shielding gas, and for carbon steels using CO2. CO2 and argon-CO2 mixtures are known as metal active gas (MAG) processes. MIG is an attractive choice to replace MMA, offering high deposition rates and high productivity.

GMAW is another form of joining that uses a consumable electrode to heat the base metal and the electrode itself in order to join pieces of material together. This process uses an electrode that is in the form of a wire, diameter and metallic makeup will be dictated by the particular materials and work being done. It is used on all thicknesses of steels, aluminium, nickel, stainless steels etc. Due to the process being semiautomatic, low operator skill is required compared to other welding processes. It is used in most metal forming/working environments.

GMAW is also known by many names depending on the country and the region; some of them are listed below:

- GMAWGas Metal Arc Welding
- MIGMetal Inert Gas
- MAGMetal Active Gas
- CO2Carbon Dioxide + Oxygen

GMAW is an American description of the process. MIG welding means the use of an inert (i.e., non-active) gas; MAG welding requires the use of an active gas (i.e. carbon dioxide and oxygen). CO2 is a more commonly used abbreviation of MAGS welding gas.

GMAW Application

1. Car body manufacturing and repairing.
2. Exhaust system manufacturing.
3. Container and storage tank manufacturing.
4. Construction industry, structural steel works.
5. Ship building.
6. Pipe work and pluming.
7. Domestic appliance manufacturing.

Description of the Process

The process consists of a DC arc burning between a thin bare metal wire electrode and the work piece. The arc and weld area are protected by an envelope of gas shielding it from the surrounding atmosphere. The wire electrode is fed from a spool, through a welding torch which is connected to the positive terminal into the weld zone. The arc is self adjusting; any variation in the arc length made by the welder produces a change in the burn off rate of the electrode, and the arc re-establishes its original length.

Equipment

The basic equipment consists of:

- Power source: A direct current unit of design and capacity according to particular usage, i.e. motor generator for site use or a rectifier for shop use.
- Wire drive unit: Containing feed rolls connected to a variable speed motor.
- Shielding gas: Correct gas for material, thickness, etc. required.
- Torch/gun: Connected to the wire feed unit to allow both gas and wire to travel to weld area.

Electrical Conditions

Welding Current and Wire Feed Speed

These are generally set by the same control; higher wire speeds mean higher amperages and vice-versa.

Voltage

Open circuit voltages can be either varied or stepped.

Shielding Gas

In addition to general shielding of the arc and the weld pool, the shielding gas performs a number of important functions:

- forms the arc plasma
- stabilises the arc roots on the material surface
- ensures smooth transfer of molten droplets from the wire to the weld pool

Therefore, the shielding gas will have a considerable effect on the stability of the arc and metal transfer and the behaviour of the weld pool, in particular, its penetration. General purpose shielding gases for MIG welding are mixtures of argon, oxygen and C02, and special gas mixtures may contain helium. The gases which are normally used for the various materials are:

Steels

- CO2
- argon +2 to 5% oxygen
- argon +5 to 25% CO2

Non-ferrous

- argon
- argon / helium

Argon based gases, compared with CO2, are generally more tolerant to parameter settings and generate lower spatter levels with the dip transfer mode. However, there is a greater risk of lack of fusion defects because these gases are colder. As CO2 cannot be used in the open arc (pulsed or spray transfer) modes due to high back-plasma forces, argon based gases containing oxygen or CO2 are normally employed.

Metal Transference Across the Arc

Dip Transfer

Also known as short circuiting arc welding, is carried out using currents below 200 amps and 25 volts. Under these conditions the arc is so short that the molten globules at the electrode (wire) tip short circuit to the work piece at rapid time intervals. The rise in the current melts the electrode tip and this re-establishes the arc. This cycle occurs approximately 100 times per second. Dip transfer is therefore suitable for all positions welding, thin material, open butt type joint etc.

Spray Transfer

Using high currents and voltages, i.e. 250-500 amps and 25+ volts, the metal is transferred across the arc in the form of fine droplets in a spray. It is used for high deposition rates and deeper penetrating welds. Spray, due to the high current and voltages involved, is mainly used on thick materials in the flat position only (except when welding aluminium, where spray transfer is used on all positions).

Inductance Settings

This control can only be used on dip transfer, and it fine-tunes the arc by either making the arc hotter or cooler. To rapid a rise during short circuiting will cause globules to explode out of the arc at the current peak, thus creating excessive spatter. Too slow a current rise will cause the electrode to freeze in the weld pool. Therefore:

1. Low inductance setting: Gives higher short circuiting frequency and relatively cold welding
2. High inductance setting: Gives lower short circuiting frequency and longer arcing periods between short circuiting giving hotter welding.

Advantages

1. Good manual application
2. Capable of a high degree of mechanisation
3. Applicable to most coppers and copper-based alloys
4. Good mechanical joint properties
5. No flux requirements
6. Heat affected zone can be localised
7. Low distortion when properly controlled
8. Low cost filler metals

Disadvantages

1. Fusion of parent metals may create metallurgical problems and porosity in joint.
2. High degree of manual skill required.

CHAPTER 8

THE DIESEL ENGINE AND THE ENVIRONMENT: A MECHANICAL ENGINEER'S PERSPECTIVE

"There is nothing free in the world except air." We often hear this from people who complain that prices are getting high. It is ironic because air is such a vital part of the world and yet we tend to take it for granted. Have we ever stopped and actually thought of the importance of air? Only few of us will probably say yes. Air is so passive and inert that we actually forget that it exists. Of course strictly speaking we can never forget that it exists because it is the air that we breathe. We live at its expense. Instead of returning our gratitude we destroy it. We have become selfish human beings who only think about themselves. Over the years our lives got more and more complicated. Here came the Renaissance period that brought vast improvements in technology. Leonardo da Vinci appeared and introduced machines and mechanical devices that were praised as great achievements. This advancement came at the price of environment. We took up Environmental Science and we saw how troubled our air is. Its quality is fast deteriorating because of our own doing. Why did we have to let it reach this far? We cannot just sit and watch the air die because if we do, then it is we who are going to die in the end. As one mother from New York said, "Our children would like to play baseball, but they can't because they can't breathe."

The Diesel Engine

In 1892, Rudolph Diesel, a German engineer developed the "diesel" engine. Unlike the gasoline engine, the diesel engine is capable of spontaneous combustion of liquid fuel without requiring spark ignition. The chief advantage of the diesel engine over the gasoline engine is its fuel economy and durability. Known as a "workhorse" the diesel engine became the engine of choice for heavy construction machinery, mining equipment, forklifts, heavy trucks, buses and trains. And it is also now used in some automobiles.

Diesel engines convert chemical energy contained in diesel fuel into mechanical power. Under optimal conditions, fuel is self-ignited as it is injected into air that has been heated by compression and during the process emits carbon dioxide and water. However, under normal conditions, diesel engines not perfectly tuned emit harmful by-products of combustion. This waste material is made up of approximately 99 per cent of substances found in the normal atmosphere (carbon dioxide and water) and one per cent of toxic substances. It is these by-products that are hazardous to the health of workers and to the environment.

Main Difference Between Gas and Diesel Engine

There are several differences between gas and diesel engines. A diesel engine does not need spark plugs to ignite the fuel. It takes in air, compresses it then injects fuel into the compressed air. The high air temperature ignites the fuel. On the other hand, a gas engine takes in a mixture of air and gas, compresses them, and ignites them using a spark plug. Another difference is the process of injection. Most gas engines use port injection, where the air, as well as the fuel go in the cylinder during the intake stroke and then compressed. There is a limit in the compression of the air and fuel mixture because if the air is compressed too much, the mixture will naturally ignite. On the other hand, diesel engines use direct injection where the fuel is directly injected into the cylinder not until the power stroke. This is why a diesel engine compresses only air, making the compression ratio higher. Compression ratio is the ratio of the original volume and the final volume. A ratio of 2:1 means compression to half its original volume. A ratio of 3:1 signifies that air is compressed to one-third its original volume. Gas engine compression ratio differs from 6:1 to 10:1. Meanwhile, diesel engine compression ratio differs from 12:1 to 24:1. More power is generated when the compression ratio is higher. This means an improved efficiency.

Diesel-powered engines are less expensive to operate than gasoline-powered engines, partly because diesel engine burn fuel oils, which call for less refining than higher-grade fuels such as petrol. This makes them cheaper. Diesels burn less fuel and emit fewer waste products and greenhouse gases, including carbon dioxide that contribute to climate change. However, while the diesel engine emits less carbon monoxide than the gasoline engine, it is also noisier, and more odorous. It is considered the most efficient power plant among all known types of internal combustion engines.

Characteristics of Diesel Fuel

Diesel fuel is a petroleum product analogous to kerosene and home heating oil. It is safer to handle than gasoline and less expensive to produce. It is oilier and heavier, which makes it evaporate more slowly than gasoline. It has a higher boiling point than water. Because of its low volatility (the tendency to vaporize) and high flash point (temperature at which diesel fuel ignites), it is less probable for it to catch fire during an accident. Diesel fuel has a higher energy density. On the average, one gallon of diesel fuel contains more energy than one gallon of gasoline. This is the reason why diesel engines get superior mileage than equivalent gasoline engines.

Parts of a Diesel Engine

Diesel engines are similar in appearance to spark-ignition engines. The parts of a diesel engine are cylinder, fuel injector, exhaust and air-intake valve, piston, crack shaft, and crankcase. A diesel has one or more cylinder, usually four, six, or eight. Pistons inside the cylinders are connected by rods to a crankshaft. As pistons move up and down in the cylinders, they cause the crankshaft to rotate. This turns axles, causing the wheels to rotate. Diesel engines tend to be heavier and more expensive to manufacture because they compress air inside the cylinder with greater force as compared to gasoline engines.

How A Four-Cycle Diesel Engine Works

There are four strokes involved in the cycle. These are the intake stroke (first downward stroke), compression stroke (first upward stroke), power stroke (second downward stroke), and the exhaust stroke (second upward stroke). A stroke is an up or down movement of a piston. The cycle begins with the intake stroke when the piston moves down and draws air into the cylinder through the intake valve. This is then followed by the compression stroke. The piston moves up and compresses the air. This compression causes the air temperature to rise to about 480 °C and creates pressure to about 100 kilograms per square centimeter. Before this stroke is about to end, the fuel is injected into a cylinder. For the fuel and air to mix well, the fuel is injected under high pressure as a spray through the fuel injector. Combustion process usually begins at the end of the compression stroke. When fuel is injected into the cylinder, it mixes with the hot air and burns explosively. Gases produced by this combustion push the piston down for the power stroke. The piston then moves up and forces the burned gases produced by combustion out of the cylinder through the exhaust valve. This last stroke is the exhaust stroke.

Emissions of A Diesel Engine

Air pollution and global climate changes are caused by internal combustion engines. Pollution has a detrimental impact on our health and the environment. Although diesels have environmental benefits such as low greenhouse gas emissions, these are offset by growing concerns with emission of nitrogen oxides and diesel particulates.Diesel exhaust, generally known as "diesel fumes" is composed of complex mixture of thousands of chemicals, gases and fine particles, and air contaminants. Diesel exhaust in the gas phase are: water vapor, oxygen, nitrogen, carbon dioxide, carbon monoxide, nitrogen oxides, sulfur compounds, volatile organic compounds, and low-molecular-weight hydrocarbons. Toxic exhausts in the gas phase include aldehydes (formaldehyde, acetaldehyde, acrolein), benzene, nitrosamines, and polycyclic aromatic hydrocarbons. Exhausts in the particulate form include elemental carbon (soot), adsorbed organic compounds, small amounts of sulfate, nitrate, metals and other trace elements. The carbon particle or soot content varies from 60% to 80% depending on the fuel used and the type of engine. Most of the contaminants are adsorbed onto the soot.

Carbon monoxide is a chemical compound of carbon and oxygen with the formula CO. It is an odorless and colorless gas, and about 3 percent lighter than air. It is poisonous to all warm-blooded animals and also to many other forms of life. If it is inhaled by humans, it combines with hemoglobin in the blood, preventing the absorption of oxygen and resulting in asphyxiation. Carbon monoxide is formed through incomplete combustion. Automobile engines are prone to this because combustion takes place quickly. Automobile-exhaust gases contain harmful quantities of carbon monoxide, sometimes several percent, although antipollution devices are set to keep the level below 1 percent.

Nitrogen dioxide (NO2) is a suffocating, brownish gas. It belongs to a family of highly reactive gases called nitrogen oxides (NOx). These gases form when fuel is burned at high temperatures, and come principally from motor vehicle exhaust. Nitrogen dioxide is a strong oxidizing agent that reacts in the air to form corrosive nitric acid, as well as toxic organic nitrates. It also plays a major role in the atmospheric reactions that produce ground-level ozone (or smog).

Particulate matter is the term for solid or liquid particles found in the air. Some particles are large or dark enough to be seen as soot or smoke. Others are so small they can be detected only with an electron microscope. Particulate matter can be directly emitted or can be formed in the atmosphere when gaseous pollutants such as SO2 and NOx react to form fine particles. This contributes to the formation of smog.

There are several factors that affect the composition of diesel fumes. The quantity and composition may vary depending on the quality of diesel fuel used, the type of engine, e.g. standard, turbo or injector, the state of engine tuning, the fuel pump setting, the workload demand on the engine, the engine temperature, and whether the engine has been regularly maintained.

Major Air Pollutants

Sources of major air pollutants include individual actions, such as driving a car, and industrial activities, such as manufacturing products or generating electricity.

Health Effects of Diesel Engine Exhaust

Most emissions of diesel engines have harmful effects on the health and the environment. Diesel exhaust contains more than 40 toxic air contaminants. These emissions are classified as toxicants, carcinogens, reproductive and developmental hazards, and endocrine disruptors.

General health effects from short-term exposure to the emissions include:

- Irritation of the eyes, nose and throat
- Lightheadedness
- Feeling "high"
- Heartburn
- Headache
- Weakness, numbness and tingling in the extremities
- Chest tightness
- Wheezing
- Vomiting

Note: 1 cubic meter (1m3) is equal to 35.3 cu ft; 1 milligram (1 mg) is equal to 0.00004 oz; 1 microgram (1µg) is equal to 0.00000004 oz.

Pollutants	Major Sources	Comments
Carbon monoxide (CO)	Motor-vehicle exhaust; some industrial processes	Health standard: 10 mg/m3 (9 ppm) over 8 hr; 40 mg/m3 over 1 hr (35 ppm)
Sulfur dioxide (SO2)	Heat and power generation facilities	Health standard: 80 µg/m3 (0.03 ppm) over a year; 365 µg m3 over 24 hr (0.14 ppm)
Particulate matter	Car exhaust; industrial processes; refuse incineration; heat and power generation	Health standard: 50 µg/m3 over a year; 150 µg m3 over 24 hr; composed of carbon, nitrates, sulfates, and many metals including lead, copper, iron, and zinc
Lead (Pb)	Motor-vehicle exhaust; lead smelters; battery plants	Health standard: 1.5 µg/m3 over 3 months
Nitrogen dioxide (NO2)	Motor-vehicle exhaust; heat and power generation; nitric acid; explosives; fertilizer plants	Health standard: 100 µg/m3 (0.05 ppm) over a year; reacts with hydrocarbons and sunlight to form photochemical oxidants
Ozone (O3)	Formed in the atmosphere by nitrogen oxides, hydrocarbons, and sunlight	Health standard: 235 µg/m3 (0.12 ppm) over 1 hr

As little as 1/1000 of 1 percent of carbon monoxide present in the air can cause symptoms of poisoning and may prove lethal in less than 30 minutes. Carbon monoxide is a major component of the air pollution in urban areas. Since it is odorless, carbon monoxide is a subtle poison. It causes only mild symptoms of headache, nausea, or fatigue, followed by unconsciousness.

Fine particulate matter can cause cancer, damage to lung tissue, cause premature death, and increase respiratory illness. The elderly, children, and people with chronic lung disease, influenza, or asthma, are especially sensitive to the effects of particulate matter. In 1996, EPA expanded the particle standard to PM 2.5 to include smaller particles. The new standard focuses on smaller particles that are likely responsible for adverse health effects because of their ability to reach the lower regions of the respiratory tract. EPA's health-based national air quality standard for PM-2.5 is 15 μg/m3 measured as an arithmetic mean and 65 μg/m3 24-hour average.

Among the health effects associated with diesel particulates:

- Asthma symptoms have been positively linked to truck exhaust (Duhme and Weiland, 1996).
- The proximity of schools to freeways and major truck routes is significantly associated with chronic respiratory symptoms in children (Speizer and Ferris, 1973).
- Children living near major diesel throughfares are more likely to suffer from reduced lung function (Brunekreef, et. al., 1997).
- Particulates in diesel exhaust bind to pollen in the air, exacerbating allergies and asthma (Knox, et. al. 1997).
- EPA has found that exposure to diesel exhaust, even at low levels, is likely to pose a risk of lung cancer, as well as other respiratory risks (EPA Report, 1996)

Diesel emissions that are considered to be carcinogenic include: benzene, formaldehyde, 1,3-butadiene, and ethylene dibromide. There are at least 16 hydrocarbons classified as possibly carcinogenic to humans. In addition to containing these carcinogens, diesel exhaust contains high levels of fine soot, known as Small Particulate Matter, or SPM. The microscopic soot is easily inhaled deep into the lungs. The worst SPM are particles less than 3.5 microns in diameter. 90% of particles emitted by diesel vehicles are less than 1 micron in diameter. These particles are not easily exhaled, and stay inside the lungs for long periods of time, corroding cells and ultimately leading to tumor growth.

Technologies that Can Improve the Emissions

Now that people are more aware of the harmful effects of the diesel engine emissions, some technologies are being developed to minimize these emissions. Tighter environmental regulations and advanced emission controls are being started worldwide.

The U.S. Environmental Protection Agency (EPA) is introducing new restrictions that are intended to lessen nitrogen oxides and particulates. These will be implemented in 2004, with full compliance required by 2009. The EPA wants a 97 percent reduction in sulfur from diesel fuel by the year 2006.

In response to this, auto manufacturers are working to develop special after-treatment systems for diesel exhaust. Peugeot, a French car company has introduced a particulate trap. This device captures soot particles produced by the engine and the device burns them to levels low enough to be practically immeasurable.

In October 2002, George We and Bob Hammerle of Ford's Scientific Research Lab were given an R&D 100 Award, also known as "an Oscar of Invention." They developed Ford Motor Company's On-board Reductant Delivery (ORD) system. This exhaust treatment system is aimed to help meet strict future diesel emissions standards worldwide, including European Stage IV and V limits for smog-forming emissions. The ORD system can be easily fitted onto existing diesel vehicles for approximately one-tenth of the cost of existing systems. The system is an air-assisted reductant injection system. Reducing agents are required by almost all diesel after-treatment devices dealing with oxides of nitrogen (NOx) and particulate mater (PM) to enhance the efficiency and durability of the device. Diesel fuel or urea water solutions can act as reductants. To be effective, these reductants must be added quickly and delivered in precisely measured and finely dispersed quantities. In addition, the delivery system must be able to withstand the high temperatures and harsh environment surrounding the automobile exhaust system. The Ford ORD system can supply any liquid form of reductant at any location needed for these applications. Ford's ORD system has only three components: an air supply, reductant injector and a reductant reservoir. It provides a more than 75 percent reduction in components, cost, size, weight, noise, and energy consumption when compared with competitor products.

Some companies are already using coconut oil-based diesel. Raja Ferry Co. Ltd, a Thailand-based fleet owner, is the pioneer of the use of coconut oil as fuel for its boats. Coco diesel is more environment-friendly than regular diesel and it can help improve the falling coconut industry. It is cheaper and almost does not contribute to pollution. In the Philippines, the Department of Agriculture, through the Philippine Coconut Authority (PCA), is persisting for the use of coco diesel. It has a sulfur content of only 0.05 percent while the regular diesel has 0.2 percent. Coco diesel will be able to comply with the sulfur content regulation for engine fuel for the year 2004 by the Republic Act of Philippines No. 8749, or the Clean Air Act. The Research and Development group of PCA is doing a research on how to produce a bio-diesel fuel that is cheap and can be practically be used in motor vehicles.

Another development in the diesel fuel involves a cleaner burning diesel fuel developed by ARCO. ARCO is a company based in Los Angeles that produces petroleum products. Four years ago it announced to have developed a new fuel, called Emission Control Diesel, which reduced soot emissions from buses, trucks and cars. During the preliminary testing of this fuel, there was a 5 percent reduction in nitrogen oxide emissions and a 15 percent reduction in the particles that include soot, with no loss in fuel economy. According to ARCO spokesman Paul Langland, the benefits of the fuel are greatly increased when it is combined with a catalytic converter, a chemical filter system often attached to mufflers that help reduce air pollution. The combination is up to 90 percent reduction in emissions. This fuel is \$0.10- \$0.15 above the cost of conventional diesel fuel.

Conclusion

It is impossible to predict what the world will look like in the year 2050. However, by looking at the past and observing trends in society, it is possible to gain a level of insight in the future that is relevant for today's decision making. New developments in diesel engine technology that are aimed at reducing harmful emission components while maintaining a high level of energy efficiency, follow each other at an increasing quick pace. The diesel (compression ignition) engine has shown to be a reliable, energy efficient and cost effective power source for road vehicles (trucks, buses and cars) and mobile machinery (like agricultural tractors, forklift trucks and earth moving equipment). Increasingly it is clear that the inherent benefits of diesel technology—efficiency, durability and safety—have lead to diesel's emergence as playing a key role in meeting future energy and environmental goals. With new innovations and continuous improvements being done to the diesel engine, the future of diesel looks bright and its role in society definitely secure.

CHAPTER 9
THE DRILLING BURR EXPERT SYSTEM

The formation of burrs is a significantly undesirable by product of drilling operations in manufacturing plants. A burr, which is the rough edge that remains on a work piece after drilling, is formed whenever a drill bit approaches the exit surface of a work piece, appearing as a short mound of material. Because of their uneven shape, burrs cause a multitude of problems in manufacturing. For example, burrs not only cause jamming and misalignment during assembly by interfering with the component parts, but also reduce the life of these parts by promoting crack initiation. Furthermore, the sharp edges of these burrs present a safety hazard to the personnel handling these parts. For these reasons, it is necessary to include a process of removing burrs as part of drilling operations. This process is known as deburring.

However, it is only feasible to incorporate this deburring process only if its inclusion as protocol into drilling operations does not hamper the cost effectiveness of the entire drilling operation. The measure of cost effectiveness is represented by the total cost of a drilling operation that includes a deburring process, which is the sum of the cost of drilling and the cost of deburring.

According to a study, both the cost of drilling and the cost of deburring are affected by the feed rate and the rotational speed of the drill bit used in the drilling process, although with opposite effects. On one hand, the cost of drilling is reduced when the feed rate and rotational speed of the drill bit is increased, because this reduces the time to machine work pieces and, subsequently, overhead costs. On the other hand, the cost of deburring rises as the size of burrs produced increases. Therefore, increasing this feed rate and rotational speed increases the cost of deburring, and hence the cost of the total drilling operation, because such increases create burrs of larger volume. Hence, the optimized drilling parameters of feed rates and rotational speeds that minimize the total cost of the drilling operation are possible to find. These optimized drilling parameters may be obtained with the assistance of the Drilling Burr Expert System (DBES).

Introduction: The Drilling Burr Expert System

The Drilling Burr Expert System (DBES) is an internet-based computer application that accurately predicts the size of the burr produced for a given feed rate and rotational speed of the drill bit in a drilling operation. By accurately predicting the size of the burr formed, engineers are then able to ascertain the cost of deburring accurately. Subsequently, not only can they determine the total cost of the drilling operation, but, conversely, they can also determine the set of optimized drilling parameters when the total cost of the drilling operation is minimized after cost analysis. In this way, the DBES contributes to the cost effective implementation of a deburring process into the protocols of a drilling operation.

To operate the DBES, engineers enter all known parameters (drill geometry, material properties, feed rate, and speed of drill) into the DBES, and then retrieve thickness and height of the resulting burr from the system. After obtaining feedback from the DBES, engineers, if necessary, can establish an appropriate choice of feed rate and rotational speed of the drill bit in drilling operation. The DBES consists of two main elements: Drilling Burr Database System, DBDS, and Drilling Burr Control Chart, DBCC. Having discussed the cost minimization in drilling operation and the description of the DBES, this report will discuss the elements, principles, and weaknesses of the DBES.

Elements of the Drilling Burr Expert System

The Drilling Burr Database System

The Deburring Burr Database System is a database of experimental data and results on the thickness and height of the burr formed using a variety of drilling parameters, performed under laboratory conditions. The table below lists the categories of conditions that affect burr formation and their corresponding parameters. Usually the drill geometry and material properties are fixed once both the drill and work piece materials are determined. However, the process conditions, which is the third category on the table, are adjustable to minimize burr size. Once, the burr data are collected and stored on the computer database, engineers can then access this database through web browsers.

Parameters Affecting Drilling Burr Formation

Category	Parameters
Drill Geometry	Point geometry, point angel, helix angle, lip relief angle
Material Properties	Hardness, strain-hardening characteristics, ductility, tensile toughness, temperature dependence of properties
Process Conditions	Cutting speed, feed, use of coolant, use of packing
Others	Tool wear, tool material

The Drilling Burr Control Chart

The second element of the Drilling Burr Expert System is the Drilling Burr Control Chart. Unlike the DBDS, which contains experimentally determined data and results, the DBCC is an analytical tool that relates the thickness and height of the burr to the drilling parameters of feed rate and rotational speed and classifies the burrs predicted into types. Through a user-friendly graphical interface, the user enters the drilling parameters into the system. Then, the system responds with analytically predicted values of burr height and thickness, while also displaying a graphical representation of the burr. In addition, a point that corresponds to the predicted burr is marked on the Control Chart, which typecasts burrs based on parameters derived from both feed rate and rotational speed. These types will be discussed in the proceeding section of this report.

This system has two advantages over the DBDS. Firstly, it allows its user to exploit any range of feed rate and rotational speed beyond that provided by the DBDS to predict the size of burrs. Secondly, it allows its user to reverse engineer by predicting the drilling parameters of feed rate and rotational speed from a user-defined height and thickness of a burr.

Principles

The DBDS provides the empirical foundations upon which the analytical framework of the DBCC is based upon. Before further elaboration, it is necessary to explain that burrs are classified into types based on their size. For example, burrs obtained from drilling stainless steel, AISI304L can be classified into three types. The first type, burr type I, has a relatively small and uniform burr height and thickness around the hole. The burr type II is slightly larger uniform burr than the first type. The last type, burr type III, which is also called crown burr, has a larger and irregular height distribution. After classifying the burrs into types, Kim suggests that two parameters, Fn and S, may be used to map the distribution of burrs across types. The two parameters Fn and S are derived from feed rate and rotational speed respectively. The ensuing graphical representation of this distribution is known as the distribution chart.

Subsequently, an analytical model may be created from the distribution chart by defining the graphical boundaries of the empirical data that are classified corresponding to these types. Specifically, a boundary trend line may be defined between the adjacent regions on the Control Chart occupied by empirical data that are classified according to typology, by applying least-square approximation to the data involved. Therefore, in addition to predicting the size of burrs, the DBCC is also able to classify the type of burrs produced based on a given set of drilling parameters. The productivity of the drilling process is maintained so that the drilling cost, subsequently the overhead cost does not increase much while the size of burr decreases.

Experimental Validation

To establish the analytical model behind the control chart, some experimental data need to be collected. However, because the DBDS is just recently developed, the DBDS only includes one material, which is stainless steel, AISI304L. Stainless steel is selected as the work piece material because it is widely used in industry and also has high modulus of elasticity and ductility. The modulus of elasticity represents the ability of the material to restore back to its original size after being deformed under a certain load. However, if the load exceeds its elastic limit, the material does not restore back to its original condition and stays in the plastic region. Furthermore, if the load is further increased, the material will break. The ability of the material to be plastically deformed, before it breaks, is quantified by the ductility. Therefore, the burrs resulted from deforming materials that have high modulus of elasticity and ductility, such as stainless steel, are easy to observe and classified into their types.

In the experiment, the split point twist drills are selected because primarily they have the self-centering characteristic, which reduce walking or skidding on the drill surface. Additionally, split point drills also help breakup chips, so that it prevents the chip from curling around drill spindles or motors, which can potentially cause damage or injury.

Several methods in measuring burr height and thickness are suggested, such as contact method, optical microscope method, optimal CMM method and rubber casting method. However, the most commonly used method to measure the burr height and thickness employs an optical microscope. After the drilling hole is located under the microscope using a proper magnification, the microscope is focused and centered at the inner edge of the burr. Subsequently, the dial of the microscope that controls microscope positioning is rotated so that the outer edge of the burr, where the root of the burr meets the undeformed work piece material, is focused at the center of the microscope. The distance for microscope to travel from the inner edge of the burr to the outer edge of the burr, which can be obtained from the dial reading, represents the burr thickness. Subsequently, the thickness of the burr formed and the corresponding drilling parameters of feed rate and rotational speed are recorded and stored into the database.

Weaknesses

Despite of its usefulness in the prediction of burr size, the DBES has two main weaknesses. The first weakness is that constructing an accurate analytical model requires collecting large number of data, which is time-consuming

The weaknesses of the DBES are that the process of collecting data for the database system is time-consuming and the DBCC predictions of burr type that are close to boundary lines in the control chart are not accurate. Since there are many possible combinations of drill geometry, material properties, and process conditions, many experiments are needed to construct the database system. Also, the predictions of burr type that are very close to boundary line are not accurate because the least square method that derives the boundary line is just an approximation. Therefore, in the area that is very close to the boundary line, there exist probabilities to have burr of type that is to the left or to the right of the boundary line.

Summary

In summary, this chapter has introduced the Deburring Burr Expert System: its description, elements, principles, and weaknesses. The Drilling Burr Expert System (DBES) advises engineers on the feed and rotational speed of the drill in order to minimize burr size. The first element, the DBDS is a database system containing burr experimental data. The second element, the DBCC is a tool that predicts burr thickness and height analytically and can also be used to select the feed and rotational speed of a drill to create a preferred type of burr for an application. The DBCC is connected to the DBDS so that the DBCC has a reference to create the analytical model. The weaknesses of the DBES are that the process of collecting data for the database system is time-consuming and the DBCC predictions of burr type that are close to the boundaries.

CHAPTER 10
BECOMING A MECHANICAL ENGINEER

The field of mechanical engineering is normally considered the broadest of all engineering disciplines. The work of mechanical engineering ranges from the ocean bottoms to space.

Education

Degrees in mechanical engineering are offered at universities worldwide. In China, India, and North America, mechanical engineering programs typically take four to five years and result in a Bachelor of Science (B.Sc), Bachelor of Technology (B.Tech), Bachelor of Engineering (B.Eng), or Bachelor of Applied Science (B.A.Sc) degree, in or with emphasis in mechanical engineering. In Spain, Portugal and most of South America, where neither BSc nor BTech programs have been adopted, the formal name for the degree is "Mechanical Engineer", and the course work is based on five or six years of training.

In the U.S., most undergraduate mechanical engineering programs are accredited by the Accreditation Board for Engineering and Technology (ABET) to ensure similar course requirements and standards among universities. The ABET web site lists 276 accredited mechanical engineering programs as of June 19, 2006. Mechanical engineering programs in Canada are accredited by the Canadian Engineering Accreditation Board (CEAB), and most other countries offering engineering degrees have similar accreditation societies.

Some mechanical engineers go on to pursue a postgraduate degree such as a Master of Engineering, Master of Science, Master of Engineering Management

(MEng.Mgt or MEM), a Doctor of Philosophy in engineering (EngD, PhD) or an engineer's degree. The master's and engineer's degrees may or may not include research. The Doctor of Philosophy includes a significant research component and is often viewed as the entry point to academia.

Coursework

Standards set by each country's accreditation society are intended to provide for uniformity in fundamental subject material, promote competence among graduating engineers, and to maintain confidence in the engineering profession as a whole. Engineering programs in the U.S., for instance, are required by ABET to show that their students can "work professionally in both thermal and mechanical systems areas." The specific courses required to graduate, however, may differ from program to program. Universities will often combine multiple subjects into a single class or split a subject into multiple classes, depending on the faculty available and the university's major area(s) of research. Fundamental subjects of mechanical engineering usually include:

- statics and dynamics
- strength of materials and solid mechanics
- nstrumentation and measurement
- thermodynamics, heat transfer, energy conversion, and HVAC
- fluid mechanics and fluid dynamics
- mechanism design (including kinematicsand dynamics)
- manufacturing technology or processes
- hydraulics and pneumatics
- engineering design
- mechatronics and control theory
- drafting, CAD (usually including solid modeling), and CAM

Mechanical engineers are also expected to understand and be able to apply basic concepts from chemistry, chemical engineering, electrical engineering, civil engineering, and physics. Most mechanical engineering programs include several semesters of calculus, as well as advanced mathematical concepts which may include differential equations and partial differential equations, linear and modern algebra, and differential geometry, among others.

In addition to the core mechanical engineering curriculum, many mechanical engineering programs offer more specialized programs and classes, such as robotics, transport and logistics, cryogenics, fuel technology, automotive engineering, biomechanics, vibration, optics and others, if a separate department does not exist for these subjects.

Most mechanical engineering programs also require varying amounts of research or community projects to gain practical problem-solving experience. Mechanical engineering students usually hold one or more internships while studying, though this is not typically mandated by the university.

License

Engineers may seek license by a state, provincial, or national government. The purpose of this process is to ensure that engineers possess the necessary technical knowledge, real-world experience, and knowledge of the local legal system to practice engineering at a professional level. Once certified, the engineer is given the title of Professional Engineer (in the United States, Canada, Japan, South Korea and South Africa), Chartered Engineer (in the UK, Ireland, India and Zimbabwe), Chartered Professional Engineer (in Australia and New Zealand) or European Engineer (much of the European Union). Not all mechanical engineers choose to become licensed; those that do can be distinguished as Chartered or Professional Engineers by the post-nominal title P.E., P. Eng., or C.Eng., as in: John Doe, P.Eng.

In the U.S., to become a licensed Professional Engineer, an engineer must pass the comprehensive FE (Fundamentals of Engineering) exam, work a given number of years as an Engineering Intern (EI) or Engineer-in-Training (EIT), and finally pass the "Principles and Practice" or PE (Practicing Engineer or Professional Engineer) exams.

In the United States, the requirements and steps of this process are set forth by the National Council of Examiners for Engineering and Surveying (NCEES), a national non-profit representing all states. In the UK, current graduates require a BEng plus an appropriate masters degree or an integrated MEng degree plus a minimum of 4 years post graduate on the job competency development in order to become chartered through the Institution of Mechanical Engineers.

In most modern countries, certain engineering tasks, such as the design of bridges, electric power plants, and chemical plants, must be approved by a Professional Engineer or a Chartered Engineer. "Only a licensed engineer, for instance, may prepare, sign, seal and submit engineering plans and drawings to a public authority for approval, or to seal engineering work for public and private clients." This requirement can be written into state and provincial legislation, such as Quebec's Engineer Act. In other countries, such as Australia, no such legislation exists; however, practically all certifying bodies maintain a code of ethics independent of legislation that they expect all members to abide by or risk expulsion.

Salaries and Workforce Statistics

The total number of engineers employed in the U.S. in 2004 was roughly 1.4 million. Of these, 226,000 were mechanical engineers (15.6%), second only to civil engineers in size at 237,000 (16.4%). The total number of mechanical engineering jobs in 2004 was projected to grow 9% to 17%, with average starting salaries being $50,236 with a bachelor's degree, $59,880 with a master's degree, and $68,299 with a doctorate degree. This places mechanical engineering at 8th of 14 among engineering bachelors degrees, 4th of 11 among masters degrees, and 6th of 7 among doctorate degrees in average annual salary. The median annual income of mechanical engineers in the U.S. workforce is roughly $63,000. This number is highest when working for the government ($72,500), and lowest when doing general purpose machinery manufacturing in the private sector ($55,850).

Canadian engineers make an average of $29.83 per hour with 4% unemployed. The average for all occupations is $18.07 per hour with 7% unemployed. Twelve percent of these engineers are self-employed, and since 1997 the proportion of female engineers has risen to 6%.

Modern Tools

Many mechanical engineering companies, especially those in industrialized nations, have begun to incorporate computer-aided engineering (CAE) programs into their existing design and analysis processes, including 2D and 3D solid modeling computer-aided design (CAD). This method has many benefits, including easier and more exhaustive visualization of products, the ability to create virtual assemblies of parts, and the ease of use in designing mating interfaces and tolerances.

Other CAE programs commonly used by mechanical engineers include product lifecycle management (PLM) tools and analysis tools used to perform complex simulations. Analysis tools may be used to predict product response to expected loads, including fatigue life and manufacturability. These tools include finite element analysis (FEA), computational fluid dynamics (CFD), and computer-aided manufacturing (CAM).

Using CAE programs, a mechanical design team can quickly and cheaply iterate the design process to develop a product that better meets cost, performance, and other constraints. No physical prototype need be created until the design nears completion, allowing hundreds or thousands of designs to be evaluated, instead of a relative few. In addition, CAE analysis programs can model complicated physical phenomena which cannot be solved by hand, such as viscoelasticity, complex contact between mating parts, or non-Newtonian flows

As mechanical engineering begins to merge with other disciplines, as seen in mechatronics, multidisciplinary design optimization (MDO) is being used with other CAE programs to automate and improve the iterative design process. MDO tools wrap around existing CAE processes, allowing product evaluation to continue even after the analyst goes home for the day. They also utilize sophisticated optimization algorithms to more intelligently explore possible designs, often finding better, innovative solutions to difficult multidisciplinary design problems.

Subdisciplines

The field of mechanical engineering can be thought of as a collection of many mechanical disciplines. Several of these subdisciplines which are typically taught at the undergraduate level are listed below, with a brief explanation and the most common application of each. Some of these subdisciplines are unique to mechanical engineering, while others are a combination of mechanical engineering and one or more other disciplines. Most work that a mechanical engineer does uses skills and techniques from several of these subdisciplines, as well as specialized subdisciplines. Specialized subdisciplines, as used in this article, are more likely to be the subject of graduate studies or on-the-job training than undergraduate research. Several specialized subdisciplines are discussed at the end of this section.

Frontiers of Research

Mechanical engineers are constantly pushing the boundaries of what is physically possible in order to produce safer, cheaper, and more efficient machines and mechanical systems. Some technologies at the cutting edge of mechanical engineering are listed below.

Composites

Composites or composite materials are a combination of materials which provide different physical characteristics than either material separately. Composite material research within mechanical engineering typically focuses on designing (and, subsequently, finding applications for) stronger or more rigid materials while attempting to reduce weight, susceptibility to corrosion, and other undesirable factors. Carbon fiber reinforced composites, for instance, have been used in such diverse applications as spacecraft and fishing rods.

Mechatronics

Mechatronics is the synergistic combination of mechanical engineering, electronic engineering, and software engineering. The purpose of this interdisciplinary engineering field is the study of automata from an engineering perspective and serves the purposes of controlling advanced hybrid systems.

Nanotechnology

At the smallest scales, mechanical engineering becomes nanotechnology and molecular engineering—one speculative goal of which is to create a molecular assembler to build molecules and materials via mechanosynthesis. For now this goal remains within exploratory engineering.

Finite Element Analysis

This field is not new, as the basis of Finite Element Analysis (FEA) or Finite Element Method (FEM) dates back to 1941. But evolution of computers has made FEM a viable option for analysis of structural problems. Many commercial codes such as ANSYS, Nastran and ABAQUS are widely used in industry for research and design of components.

Other techniques such as Finite Difference Method (FDM) and Finite Volume Method (FVM) are employed to solve problems relating heat and mass transfer, fluid flows, fluid surface interaction, etc.

APPENDIX

CURRENT RESEARCH IN MECHANICAL ENGINEERING

Film Thickness Analysis for EHL Contacts under Steady-State and Transient Conditions by Automatic Digital Image Processing

E. Ciulli, T. Draexl, and K. Stadler

ABSTRACT

The knowledge of the film thickness values is very important in lubricated contacts to verify the lubrication conditions. Optical interferometry is one of the most used methodologies for film thickness and shape determination of Elastohydrodynamic-lubricated contacts. An image processing programme has been developed for the analysis of white light interferograms. The programme, based on the analysis of the hue channel, has been developed in order to process big amounts of images, as often generated under transient conditions. The measurement range is currently limited to a maximum film thickness of 0. 7 μm. The programme has been used for analysing several images recorded in tests carried out with a ball on disc contact under steady-

state as well as transient conditions for different velocities and slide-to-roll ratios. Under transient conditions, the entraining velocity is varied with a sinusoidal law for two different frequencies. The results obtained evidenced an asymmetric reduction of the film thickness when increasing the percentage of sliding, both for stationary and transient conditions. Under transient conditions with increasing test frequency, film thickness loops of increasing amplitude have been found that reduce their amplitude more when the ball is running faster than the disc. Squeeze and thermal effects can explain the results obtained.

Introduction

Film thickness measurements of Elastohydrodynamic (EHD) lubricated contacts are important to evaluate the lubrication regime. One of the most used methods of film thickness investigation is optical interferometry. This method has been successfully applied to EHL contacts since the 1960s and is particularly suitable for determining also the shape of an EHL contact. The first investigations were made under pure sliding and pure rolling conditions using experimental rigs with contacts between balls and plates [1, 2].

Some of the typical EHD-lubricated machine elements, as rolling element bearings, cams or gears, work under conditions where thermal effects due to sliding are not negligible. Furthermore, it must be considered that machine elements are working mainly under transient conditions by means of changing radius of curvature, load, or speed. Therefore, experimental investigations have been successively made including sliding effects by testing different values of the slide-to-roll ratio S or by testing under transient conditions. Just to mention some of these studies, Smeeth and Spikes [3] investigated the behaviour of S on an EHL point contact for different oils by analysing the film thickness, and Sugimura et al. [4] investigated the effect of squeeze on the film thickness under transient conditions (but the effect of S was not included).

More recently, the authors investigated nonconformal lubricated contacts between steel balls and glass discs for different S also under transient conditions [5, 6]. These previous works were more addressed to friction investigation and showed the need of further investigations and a better film thickness analysis. Particularly under transient conditions, a huge amount of interference images were recorded that could not be analysed efficiently by eye and an automatic method was considered to be necessary.

Some programmes for almost automatic analysis of monochromatic interference images have been already developed, as reported in [7, 8]. However, for reasons of a better resolution and higher light intensity, related to the reduced exposure time necessary for recording images under transient conditions, the use of interference filters for monochromatic interferometry is not preferable; white light interference, producing colour images, was therefore used for this kind of tests.

In this work, the main aspects of the programme developed capable of determining the local lubricant film thickness in the entire area of an EHL point contact are described. The programme can process large series of images with a moderate amount of user inputs. For the automatic analysis of white light interferograms different methods exist. One recent method is based on the analysis of the RGB colourspace [9], whereas the other method uses the HSV colour space [10]. The latter image processing method first introduced by Marklund is used due to its greater simplicity and minor sensitivity to illumination conditions. The programme developed is tested by analysing interference images obtained in tests carried out under conditions similar to previous works[5, 6]. The film thickness results obtained both under stationary and transient conditions for different S are presented and analysed.

Processing of White Light Interferograms

Basic Aspects of Optical Interferometry Applied to EHL Contacts

As well known, for the investigation of EHL contacts, monochromatic and white light interferometry is often used. In both cases images with fringes are formed by the interference of the beams reflected by the two surfaces of the contacting bodies separated by the lubricant. Images with dark and bright fringes are formed by the light of one determined wavelength (monochromatic interferometry) whereas a spectrum of coloured fringes is formed by white light interferometry.

The basic principle of the formation of interference patterns for monochromatic interferometry is shown in Figure 1.

In common EHL investigations, a glass disc is used coated with a semireflective chromium layer on the side of the contact. A beam of light falling through the glass will be partially reflected at the interface between the glass and the chromium layer, whereas another part will go through it and is reflected at the surface of the specimen. The silica (SiO2) layer shown in Figure 1 serves as protection to reduce abrasion and also as a spacer layer to extent the lower measurement range. Since the two beams cover different distances, there will be a phase shift between the corresponding light waves. When the waves superimpose, the resulting amplitude will be between zero (destructive interference) and twice the amplitude of the original wave (constructive interference), depending on the optical retardation. Thus, an interference pattern consisting of bright and dark fringes can be made visible in the monochromatic case and multicolour fringes using white light interferometry.

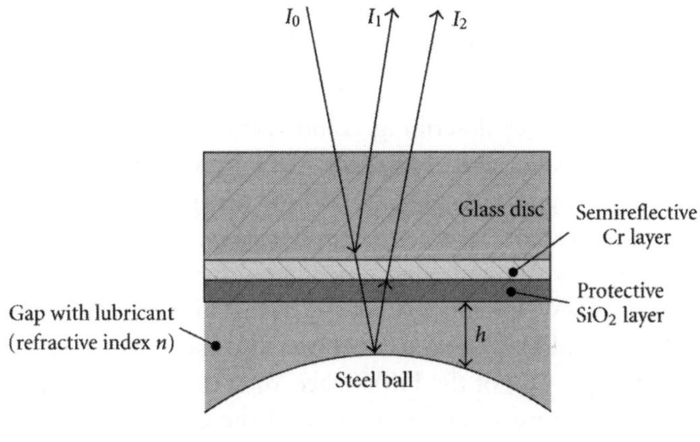

Figure 1: Basic principle of interferometric film thickness measurement.

Film thickness values are then evaluated when using monochromatic interferometry knowing the wavelength of the light and the refractive index of the lubricant considering the fringes order. The resolution is limited since normally only discrete film thicknesses at the bright and dark extremes can be calculated and the values in between must be interpolated. The distance of these discrete values for the commonly used light's wavelength and lubricant's refractive index is in the order of 90 nm.

In white light interferometry, every coloured fringe corresponds to a certain film thickness and this usually provides a better resolution than monochromatic interferometry. Considering that the correlation between film thickness and any numerical representation of colours is generally nonlinear, film thicknesses can be determined by a table lookup once an appropriate calibration table is available. The greatest disadvantage of white light interferometry is a rather strict limitation of the measurement range. Film thicknesses smaller than about 0. 1 μm tend to appear as shades of grey (with zero theoretically resulting in black), and above roughly 1 μm fringes are not visible at all due to the low-coherent nature of white light. The lower limit can be extended by adding a spacer layer of transparent material to the disc (often referred to as ultra thin interferometry), but this implicates an equivalent reduction of the upper limit. Although the measurement range is rather narrow, it is in the scale of common EHL film thicknesses.

First methods using white light interferometry are based on the human ability to compare the colours with a calibration table. It should be noted that colour perception may vary from person to person and thus "manual" analysis of white light interferograms should be taken with care and needs a lot of time and experience.

Successively, some methods have been developed to correlate the colour information stored in the recorded images to the film thickness in more or less automatic ways.

There are several ways of describing colour information quantitatively for processing on computers. RGB colour space is the most common one, since it represents each colour by its red, green, and blue component. Another colour space, which is to some extent closer to the human perception of colour, is known as HSV. Identical or very similar colour spaces are sometimes also referred to as HSB, HSI, HSL, or ⊠-S-W. It defines colour by its hue H, saturation S, and value V components. RGB and HSV can be converted into each other by a nonlinear, bijective transformation. Out of the HSV space, only the H channel is used due to its insensitivity to luminance and saturation of the colours, which are stored in the two other channels. Therefore, the measurement results are unlikely to be influenced by illumination conditions. In contrast to H, S, and V will change in a very irregular manner within a comparatively small bandwidth and therefore their analysis does not seem eligible for the measurements [11].

Full details of methods on how to use HSV and the H channel in particular for film thickness measurements can be found in [10–13]. Here only the main aspects are reported for understanding the methodologies used in the programme specifically realised for fast processing of image series from experiments under transient conditions.

The value of the H channel can be obtained from R, G, and B [10] by the equation:

$$H = \arctan 2\left(\frac{2G - R - B}{R - B}\right),, \quad (1)$$

where arctan2 is the four-quadrant inverse tangent, which requires a case differentiation depending on the quadrant of the argument.

In a white light interferogram, the progression of the H value over film thickness will be a wrapped signal similar to a periodical signal, but not exactly periodical. Since it can be regarded as a phase angle modulo 2π, it will jump from its minimum to its maximum value or vice versa. A function bounded to the interval (0, 2π) can be considered as a wrapped phase signal if it has a progression as shown in Figure 2. In this example, the hue values have been actually calculated using three wavelengths (λ1=505 nm, λ2=540 nm, and λ3=580 nm); Hvalues have been normalised to (0, 1).

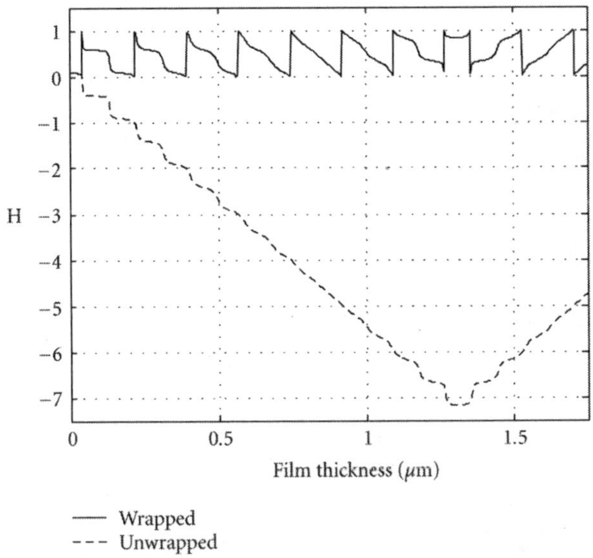

Figure 2: Example of phase wrapped and unwrapped hue values as a function of film thickness.

If a way is found to deal with the ambiguity of the H values, a calibration table correlating them directly to film thicknesses can be created. This is done by an algorithm that adds (or subtracts, respectively) the height of the interval to all values subsequent to a discontinuity, so that a continuous signal is reconstructed. This process, which is referred to as phase unwrapping, can also be accomplished by computing the numerical derivative (or gradient for phase maps), setting it to zero at the peaks resulting from the discontinuities and reintegrating it. Figure 2 also shows a characteristic behaviour of the unwrapped hue value as a function of the film thickness. The hue value monotonically decreases until a minimum, the position of which depends on the spectrum of the light. For trichromatic light, as assumed in this example, the minimum will usually be above one micron, whereas for white light it is expected between 0. 6 and 0. 8 microns with a similar course of the function as shown here. As soon as the hue values are available as a monotonic functions in this interval, a calibration table associating them uniquely to film thicknesses can be built straightforwardly. The monotonically increasing section beyond the minimum can only be used for calibration if an additional case differentiation is made.

The hue values in interferograms with repetitive coloured fringes are obviously such functions in two dimensions that will be called wrapped and unwrapped phase maps. An example of a wrapped phase map extracted from a white light interferogram of a real EHL contact is shown in Figure 3(a).

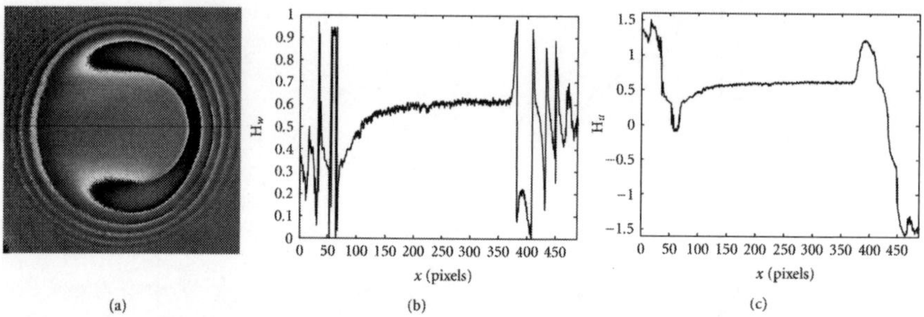

Figure 3: Wrapped phase map of a white light interferogram of an EHL point contact (a) and wrapped (b) and unwrapped (c) hue values along the horizontal line.

The wrapped hue values in Figure 3 are created by measurements and not by artificially wrapping an unbounded signal as seen in Figure 2. Measurements imply a certain amount of noise and uncertainty. In addition, if the sampling rate is too low, it is possible that the dynamics of the jumps are not fully captured. Therefore, the real jump discontinuities will not be exactly 1 but slightly smaller. Taking into account that the fringes tend to attenuate at large film thicknesses, these effects can become very distinct. Figure 3(b) shows a horizontal section through the wrapped phase map (H value) of Figure 3(a). It is clearly visible that the signal-to-noise ratio and the dynamic range deteriorate towards the edges of the interferogram. Unwrapping such a signal can be challenging, since a discrete signal by definition has discontinuities everywhere. Consequently, it must be decided which of them are caused by actual phase jumps and which by the natural course of the signal, including noise. Low-pass filtering does not solve this problem, since it both reduces noise and smears out the phase jumps, which are high-frequency components themselves. The (partially) unwrapped phase in this particular example is shown in Figure 3(c). The decision criterion (threshold) used in this work has been a discontinuity of more than 0. 3 between two consecutive values. In the result, the approximate shape of the typical mid-plane profile of an EHD contact becomes visible (however, in this particular example, the algorithm failed to perform correct unwrapping of the values near the left edge of the plot).

For measured phase maps in two dimensions another important problem arises. Line integrals over them will in general be path dependent and errors will propagate, making it difficult to obtain an unwrapped result that is consistent everywhere. Since multidimensional phase unwrapping is a task that occurs in various applications, such as magnetic resonance imaging or radar interferometry, various rather complex algorithms have been developed for it in the past. For example, the algorithm presented in [14] had received particular attention and was an important basis for the further developments conducted in [10, 12]. Several versions of the algorithms found in the literature have been implemented in the Matlab environment and tested with interference images obtained with the experimental apparatus described in the following (e. g. , [15]). However, none of the algorithms performed satisfactorily on the real interferograms of EHL contacts used. Therefore, a pseudo-two-dimensional integration approach has been finally chosen taking into account that a path-following integration method is likely to produce less inconsistencies if it uses several separate paths that all start at the same point. The two-dimensional unwrapped phase map is assembled by joining the results of all the one-dimensional integrations and interpolating the values that were not covered by any path. Since a method especially for circular point contacts was to be developed, the central point of the contact area suggested itself as the origin for integration paths running radially outwards. In order to obtain a more concise, easy to handle programme code, the algorithm does not operate directly on the original wrapped phase map. In a preprocessing step, the matrix is read out along a sequence of radial lines at intervals of a predefined angular increment. The values retrieved along each of the lines are written into rows of a new matrix. Thus the first column of this matrix will contain the wrapped value of the central point in every row, while the last column will contain the values occurring around the whole circumference of a circle with the maximum radius considered. This transformation is illustrated in Figure 4. Unwrapping itself is accomplished by a simple one-dimensional algorithm that unwraps the preprocessed matrix row by row and decides where to add or subtract integer values (because all wrapped $H \in (0,)$) by comparing the differences between consecutive elements to a threshold value $T<1$. The threshold value is given to the function as a parameter. Finally, the resulting unwrapped matrix is transformed back into the original coordinates.

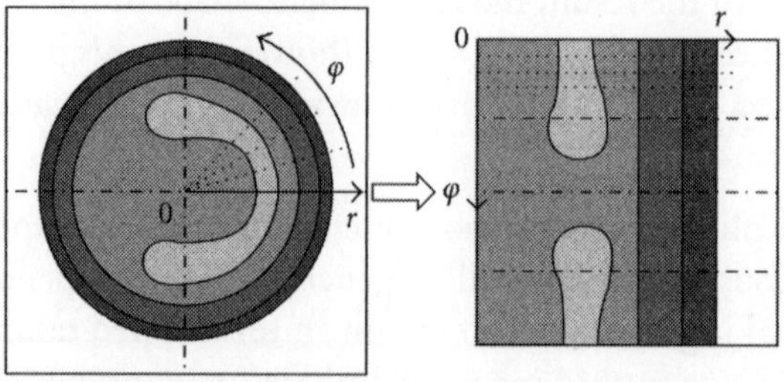

Figure 4: Schematic illustration of the coordinate transformation performed before phase unwrapping.

Filtering and interpolation operations can be included at this point to reduce the inherent noise of the measured interferograms. For all filtering operations, spatial averaging by convolution with a finite impulse response (FIR) filter was implemented. This filter replaces every matrix element by a (weighted) average of its neighbours[16]. Mathematically, this can be described by the convolution sum

$$\tilde{A}m,\ n=_{!,n} = \sum_{k}\sum_{l} \mathbf{F}_{k,l}\,\mathbf{A}_{m-k,}, \quad (2)$$

where A={Am, n} and Ã={Ãm, n}are the matrices containing the input and output values respectively, and F={fk, l} is the convolution kernel or filter mask that defines the filter. Uniform convolution kernels, consisting of an N×N matrix with each element f(k, l)=1/N2∀(k, l), turned out to deliver good results, that is, they effectively reduce noise without changing the low-frequency components of the signal.

This pseudo-two-dimensional algorithm is stable as long as the threshold value is not set smaller than twice the maximum noise amplitude that occurs in the matrix. The simple arithmetic without any calculations in the frequency domain does not cause undesirable oscillations in the result. The success of the unwrapping procedure will depend on the quality of the input image, in particular on the magnitude of the phase discontinuities and the signal-to-noise ratio.

Organisation of the Image Processing Software

For the software realised, Matlab has been chosen as the programming environment because it offers built-in functions for image and signal processing, as well as versatile data input and output capabilities.

The software follows a calibration-lookup approach. The necessary operations to obtain film thicknesses from interferograms can be split up into two main steps. First, a correlation between film thicknesses and hue values, that is, a calibration table, must be found. In a second step, the actual image analysis, this calibration table can in turn be applied to determine film thicknesses from given hue values. In both cases, phase unwrapped hue values are used. Since the calibration procedure must be accomplished at most once per experiment, the two subtasks were realised in separate programs (Figure 5). Both of them employ the same unwrapping algorithm, but differ slightly in the way they process images. This is due to the fact that for calibration the area outside of the actual (Hertzian) contact is of interest, whereas for image analysis mainly the (EHL) contact zone itself is considered.

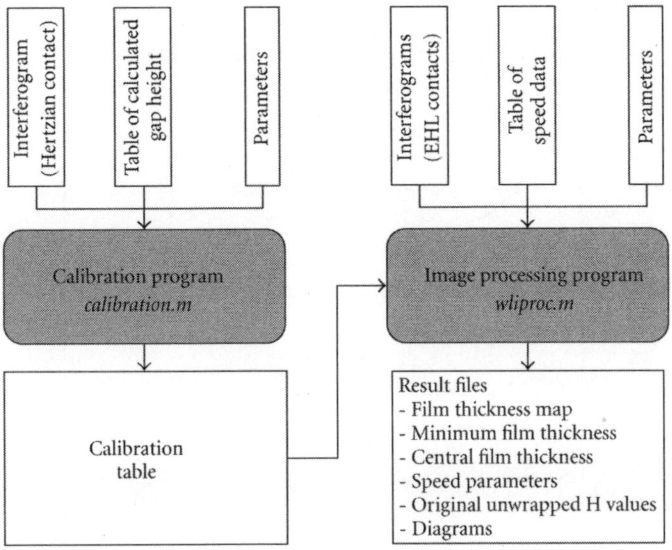

Figure 5: Structure of the two main programs and their interaction.

The calibration programme calibration. m requires as input an interferogram image of a Hertzian point contact and a table containing the height of the gap around this contact as a function of the radius around its centre. The output is a Matlab file containing the calibration lookup table. The parameters requested as user inputs include the area within the image to use and the scaling factor relating the size of a pixel in the image to the corresponding length in the real contact. In consideration of the programme being comparatively seldom used, further interaction with the operator is intended to ensure that only correctly phase unwrapped values are exploited.

In contrast, the image processing programme wliproc. m (white light interferogram processor) passes on fewer decisions to the operator. A set of parameters for each experiment must be given at the beginning. Once an image file name is provided to the programme, it looks for all files generated by the camera software. So there is no need to open every interferogram separately. This batch-processing feature is meant when processing several images in succession. A limited number of partially graphical inputs must be made until the calibration table can be applied and the result is visualised. If the user accepts it, a result file in Matlab format and several figures are created. The raw, unwrapped hue values are also saved in case the calibration table should turn out to be erroneous later.

Consideration on Accuracy

The calibration table is presented as a diagram like in Figure 6.

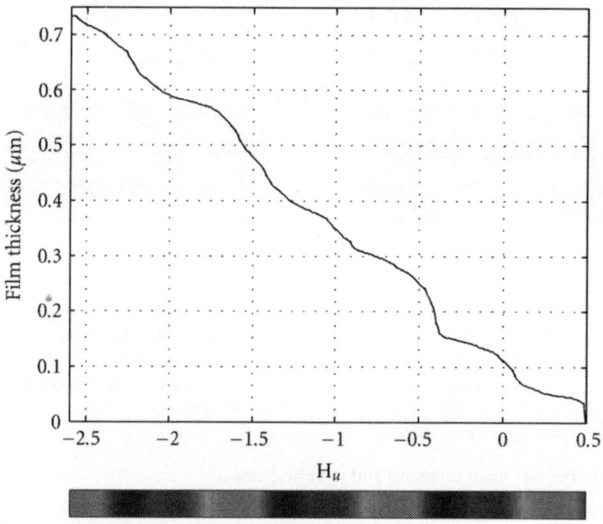

Figure 6: Calibration curve uniquely associating film thicknesses to unwrapped H values. The corresponding colours are shown in the colour bar (only hues, not considering possible variations of saturation and luminance).

The resolution of the calibration table depends on the resolution of the given hue values. For a 24-bit colour image, each channel (R, G, B or H, S, V) is represented by 8 bits and can thus adopt 255 discrete values. By phase unwrapping, which extends the width of the range of values from 1. 0 to approximately 3. 2, this number is extended to more than 800 values in the present example. They are distributed over a film thickness range of 0. 7 μm. If the calibration functions were assumed to be linear, the corresponding film thickness difference between two discriminable colours would be 0. 7 μm/800 <1 nm. Theoretically, this value could even be improved by recording and processing images of a higher-colour depth (the present camera is capable of producing 10 bits per channel). However, the influence on the resolution by other factors is significantly higher.

For example, the spatial resolution with which the interference pattern is recorded must be taken into account. The distance between two pixels in the image equals 1. 18 μm here. The steepest slope of the steel ball's contour used within the measurement range is about $\Delta h/\Delta r \approx 0.005$. Thus, the largest vertical distance between two points mapped to neighbouring pixels will be 0. 005 × 1. 18 μm ≈ 6 nm. However, for some ranges of values, the resolution might also be limited due to the nonlinear nature of the calibration function. Between 0. 16 and 0. 24 μm the curve is relatively steep, that is, a small change of the hue value corresponds to a large film thickness difference. This range is covered by roughly 20 discrete hue values, resulting in a resolution of about 4 nm. Furthermore, the shape of the Hertzian contact assumed for the calibration results from a theoretical model, whereas the real shape may deviate from the theory due to manufacturing inaccuracy and surface roughness. The surface roughness of the steel ball (mean-square-roughnessRq) was given to be 8 nm here.

Taking all these considerations into account, one can expect an uncertainty of measurement in the order of 10 nm. This assumption was tested by analysing several interferograms of different Hertzian contacts by the processing programme comparing calculated and theoretical results.

The calibration images are recorded at atmospheric pressure (the area of interest is outside the Hertzian contact), whereas the pressure in the lubricant film of an EHL contact is much higher. Moreover, pressure varies considerably within the contact area, causing locally varying deviations of the measured film thickness. These errors can be dealt with if algorithms are employed that calculate the pressure distribution from the initially measured film thicknesses. In an iterative correction scheme film thickness and pressure can then be redetermined until the procedure converges and further changes become insignificant. Such correction methods based on a pressure-density model and the Lorentz-Lorentz relation have been reported in [9, 17]. In the present version of the program, which concentrates on film thickness determination itself, a refractive index correction is not implemented yet.

In [9], a change of the film thickness after several iterations of 6% was reported for the area of the outlet constriction where the maximum pressure occurs. The authors used parameters very similar to those adopted in the present work. Therefore, the errors are assumed to lie in the same scale and thus also in the scale of the method uncertainty of measurement.

The methodology developed needs still to be improved. However, despite its limitations, a necessary important first step for the comparison between steady-state and transient conditions result is done.

Example of Programme Use

To illustrate how the film thickness is obtained from an interferogram as provided by the camera, the main steps and intermediate results are described in the following.

A sample white light interferogram has been selected that presents some light intensity differences for showing the capabilities of the programme developed (Figure 7(a)). After the image is provided as an input to the programme together with the actual pixel dimension and the calibration table, a version of the image with enhanced colours is created to facilitate the selection of the central point (Figure 7(b)).

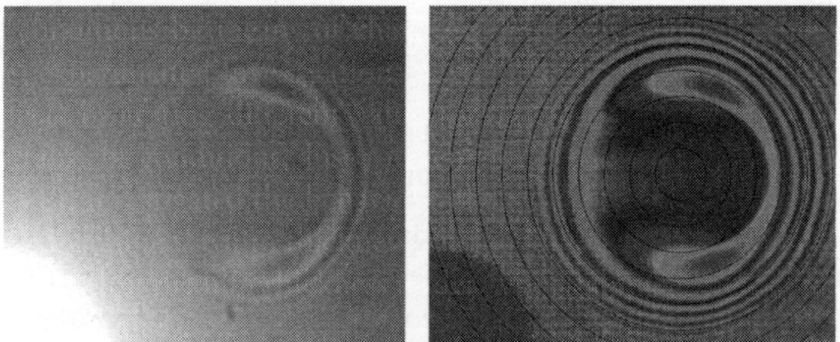

(a) (b)

Figure 7: White light interferogram of an EHL point contact as recorded by the camera (a) and image with enhanced colours and circles displayed to facilitate the selection of the central point (b).

After selecting the centre of the interferogram, phase unwrapping of the H channel is started with an initial threshold value; if results are unsatisfactory, different values can be chosen. Results obtained with an initial threshold value of 0. 25 and a successive value of 0. 2 are shown in Figure 8. A good portion of the contact zone is reproduced well in Figure 8(a) besides the peaks visible near the edge of the processed area. In this region, the dynamic range of the H signal is compressed too much to be unwrapped successfully. However, most of these points lie outside the measurement range anyway and would be excluded from further processing. A disturbing phenomenon can be observed looking at Figure 8(a), a gap appearing in one of the side lobes located close to the zone of minimum film thickness.

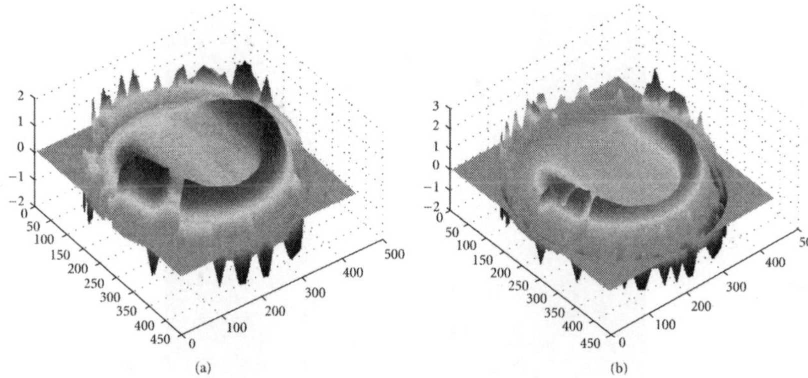

Figure 8: Phase unwrapped H channel obtained using a threshold value of 0. 25 (a) and of 0. 2 (b).

Adjusting the unwrapping threshold can improve the result (Figure 8(b)), but the error cannot be completely eliminated in this case. Thoroughly looking at the image, the reason becomes evident. Figure 7(b) shows that between the blue central zone and the green fringe in the zone of minimum film thickness there are purple, red, and yellow fringe. However, these fringes seem to be noncontinuous at the inner edge of the lower-side lobe. Here, an almost direct transition from blue to green occurs. Thus, there is no conspicuous discontinuity between the wrapped H values of neighbouring pixels, causing the unwrapping algorithm to fail. This localised absence of colours is assumed to be related to the present optical equipment.

Before offering the option to use only the flawlessly unwrapped half of the results, the programme asks the user to define up to what radius the data will be used as input for film thickness calculation. This radius is chosen by mouse click in such a way that the largest possible area is covered without including points beyond the measuring range or distortions caused by failed phase unwrapping (Figure 9). Similar to the selection of the central point, the choice is shown as a circle in the image. The last user input before actually determining film thicknesses is a mouse click on the point where minimum film thickness is assumed to be. Since the film thickness in the calibration table is a strictly monotonic function of the unwrapped hue value, this decision can already be taken before any film thicknesses are known.

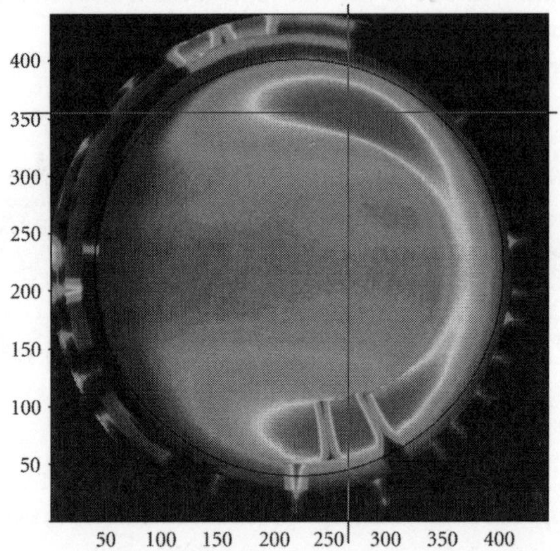

Figure 9: Selection of the useful area (black circle) and the point of minimum film thickness (cross point).

After adding a spherical shape around the utilised area and filtering the result to reduce noise, the decision to use only one half of the film thickness map can be taken. Finally, a 3D surface plot of the whole film thickness matrix (Figure 10(a)) and two 2-dimensional plots of perpendicular sections through the central point are created (Figures 10(b) and 10(c)) (the ripple in the central region could be related to the filtering level chosen for this example). This allows the user to check if the results are plausible by comparing also with a theoretical indication output obtained using standard film thickness formulas [18]. In this example, one half of the contact has been used. Finally, the result file and the plots will be saved.

Film Thickness Results

Experimental Details

The programme developed has been tested with series of images obtained with an experimental apparatus for investigation of lubricated contacts. A schematic

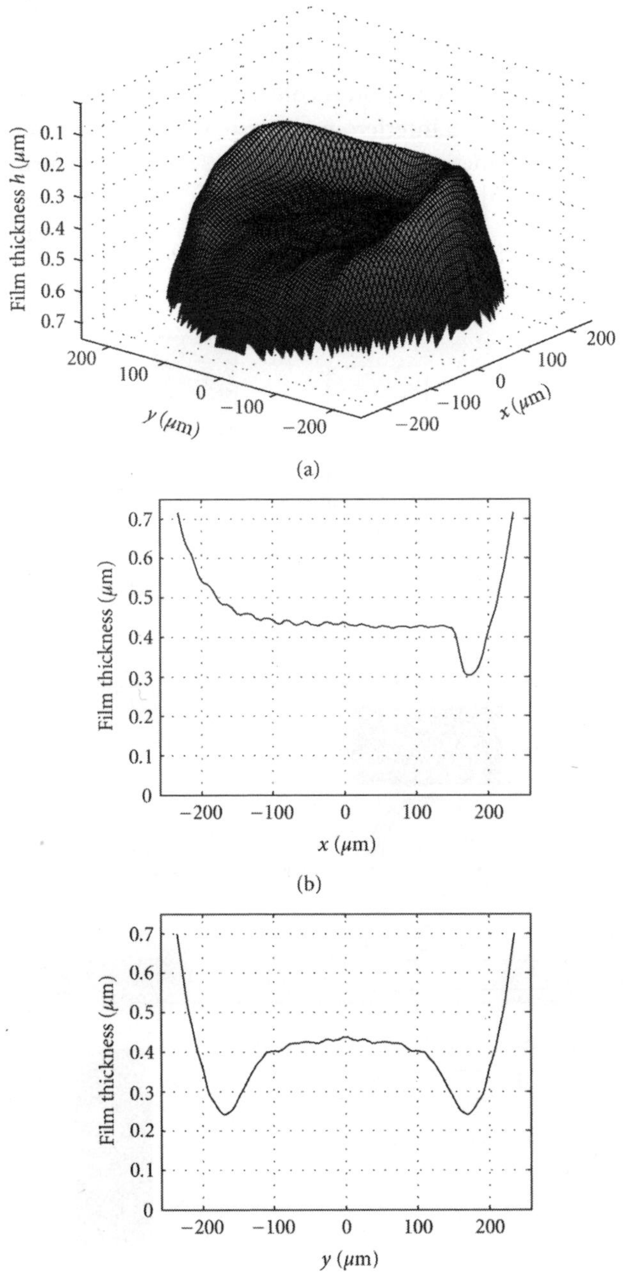

Figure 10: Final results: 3D surface plot of the film thickness (a) and film thickness in midplane sections along the x axis (entraining direction) (b) and the y axis (perpendicular to the entraining direction) (c).

drawing of the test rig used is shown in Figure 11. The specimen and the disc are driven by separate DC motors controlled through a LabView programme running on a standard PC. The load is applied via a lever mechanism supported on a radial gas bearing whose axial motion is constrained by a load cell for measuring the friction force. An optical interferometry system with white light and a high-speed camera is used to measure the film thickness and shape. Further details of the rig are reported in [5].

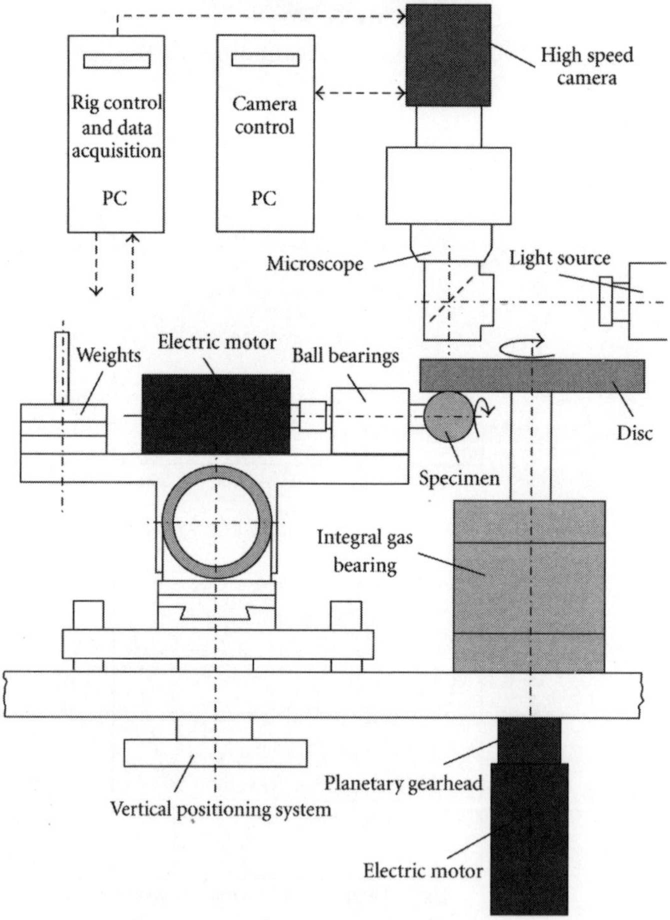

Figure 11: Schematic drawing of the test rig.

In order to investigate steady-state and transient conditions by using the programme developed for the analysis of the interference images, test conditions have been chosen producing film thickness values between 0. 04 and 0. 7 μm, roughly, the field of applicability of the program.

A glass disc with a SiO2 spacer layer of 0. 06 μm and an AISI52100 bearing steel ball with a diameter of 10. 319 mm has been used in this work. Their root mean squares roughness are Rq = 0. 01 μm and Rq = 0. 007 μm, respectively. A paraffin base mineral oil (SN500) has been used as a lubricant, whose dynamic viscosity, pressure viscosity coefficient, and density are η0=0. 497 Pa·s, α=2. 87 10–8 Pa–1, and ρ0=892 kg/m3 at test temperature and atmospheric pressure. The oil temperature and the load has been kept constant during all tests respectively at T = 12°C and F = 15 N (the corresponding Hertzian radius and pressure are 0. 1 mm and 0. 7 GPa, resp.).

Tests have been carried out under steady-state conditions for entraining velocities ue = (us + ud)/2 ranging from ue = 0. 025 to ue = 0. 25 ms–1 and different values of S = (us–ud)/ue (0, ±0. 8 and ±1. 8)(us and ud are the surface speeds of the steel ball and the disc). Under transient conditions, the entraining velocity has been varied with a sinusoidal law between umin = 0. 04 ms–1and umax = 0. 225 ms–1 for two frequencies of 0. 25 Hz and 1 Hz.

Steady-State Conditions

In Figure 12, some interferograms recorded under steady-state conditions for different values of the slide-to-roll ratio S and different velocities are shown. The oil entrainment is from the left. Such interferograms have been used as inputs of the previous described programme in order to evaluate the film thickness. However, a lot of information can be gained also by looking on the original interferograms.

As described before, in white light interferometry, the interferograms consist of multicoloured fringes with changing colours depending on the film thickness. Looking at the images of Figure 12, it can be seen that at very low entraining velocities, the interferograms are very similar in colour, independent of the slide-to-roll ratio. The horseshoe-shaped constriction at the outlet zone, which contains also the minimum film thickness, is very small and thus the colour perception is close to the central film thickness. With higher entraining velocities, the difference in colour of the horseshoe-shaped constriction and the central film thickness part is evident. Also, a clear difference can be seen between pure rolling conditions (S = 0) and rolling-sliding conditions at different S. At the highest entraining velocity and slide-to-roll ratio, also a difference in colour is evident between S = −1. 8 and S = +1. 8. These qualitative observations should be quantitatively evident in the film thickness results after analysing the interferograms by the automatic digital image processing programme.

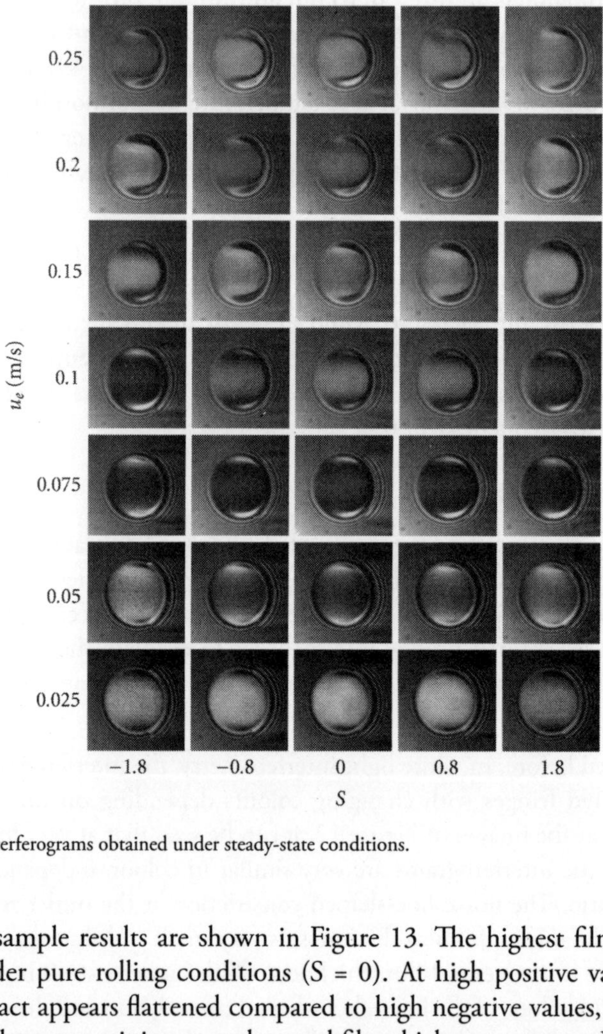

Figure 12: Interferograms obtained under steady-state conditions.

Some sample results are shown in Figure 13. The highest film thickness is found under pure rolling conditions (S = 0). At high positive values of S the EHL contact appears flattened compared to high negative values, where a large difference between minimum and central film thickness emerges. It must be annotated that this asymmetry seems to occur only in contacts between materials with different thermal properties [19].

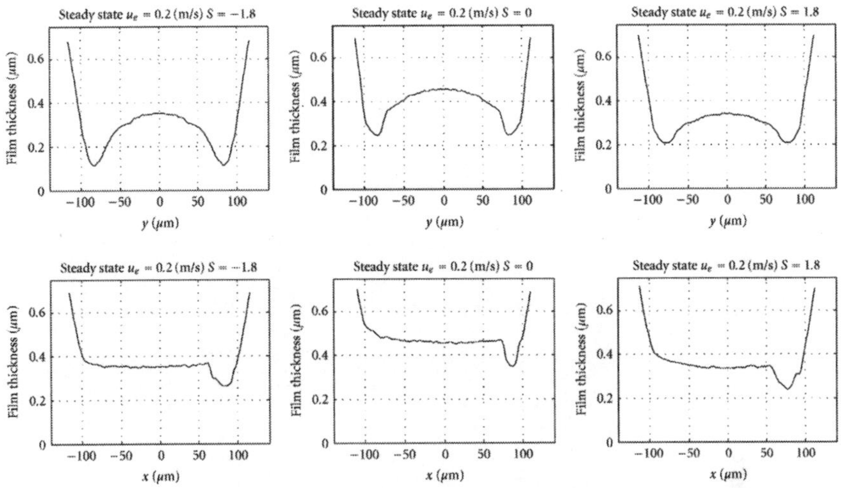

Figure 13: Film thickness profiles along the y- (top) and x-directions (bottom) (perpendicular to the entraining direction and entraining direction, resp.) under steady-state conditions for S = −1. 8, S = 0, and S = +1. 8.

The results obtained for the other values of S and ue confirm the reduction of the film thickness for S ≠ 0 compared to pure rolling conditions. In addition, it appears that the film thickness diminishes more for positive than for negative S and that these effects diminish with low entraining velocities. These results are in agreement with the behaviour found in [3].

Transient Conditions

The interferograms recorded under transient conditions at the two frequencies of 0. 25 Hz and 1 Hz are shown in Figures 14 and 15, respectively. For each value of S, the interference images recorded during one cycle are reported in two columns, starting from the highest speed at the top of the column on the left to the lowest one at the bottom (decelerating conditions, Dec), and coming back from the lowest speed to the highest one at the top of the column on the right (accelerating conditions, Acc). Only for clarity reasons, the interference images recorded at the lowest speeds are repeated twice.

Some aspects are immediately evident by the simple visual analysis of these images.

The different interferograms obtained for the same velocities under decelerating and accelerating conditions clearly indicate a shift between the speed and the film thickness variations. The differences are more evident for the highest test frequency.

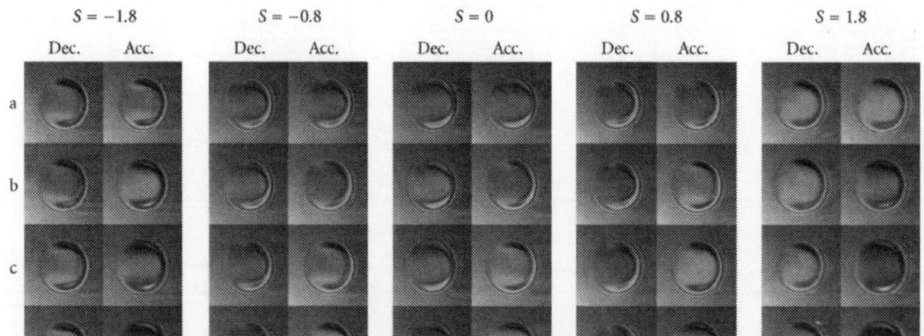

Figure 14: Interferograms obtained under transient conditions. Test frequency: 0. 25 Hz. (Entraining velocities ue: a = 0. 2250, b = 0. 2065, c = 0. 1880, d = 0. 1695, e = 0. 1510, f = 0. 1325, g = 0. 1140, h = 0. 0955, i = 0. 0770, j = 0. 0585, k = 0. 0400 m/s).

Differences are also evident between film thicknesses for positive and negative values of S as found for the steady-state conditions.

Quantitative results obtained using the programme developed are shown in Figures 16 and 17. Due to the big amount of plots obtained, results have been summarised in diagrams showing the trends of the measured values of the minimum and central film thickness for the different values of S.

Figure 16 shows the measured central film thickness, hcen, plotted as a function of time and entraining speed. The interpolated values under steady-state conditions are reported for comparison too.

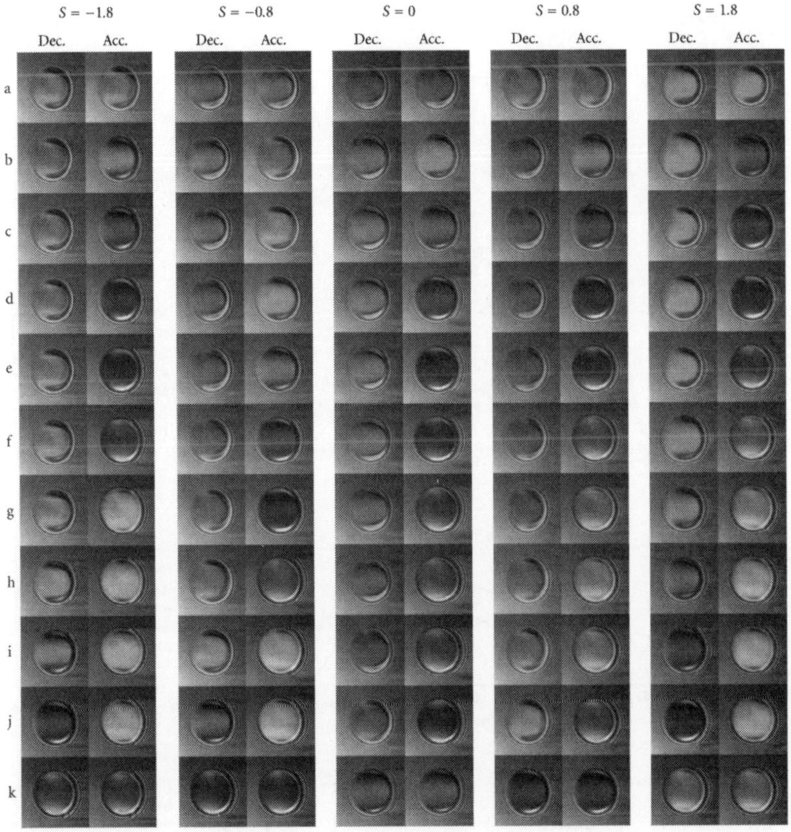

Figure 15: Interferograms obtained under transient conditions. Test frequency: 1 Hz. (Entraining velocities ue: a = 0. 2250, b = 0. 2065, c = 0. 1880, d = 0. 1695, e = 0. 1510, f = 0. 1325, g = 0. 1140, h = 0. 0955, i = 0. 0770, j = 0. 0585, k = 0. 0400 m/s).

The values under steady-state conditions were also compared with the theoretical ones showing a general good agreement for central film thicknesses corresponding to the part of the calibration curve above the step observed between 0. 16 μm and 0. 24 μm (see Figure 6). Forhcen greater than about 0. 2 μm, the slope of the hcen-velocity line in a log-log scale is about 0. 7, according to the EHL theory [18].

For an immediate comparison among the results obtained with the different values of S under transient conditions, central and minimum film thickness for the two frequencies are also shown for different values of S in Figure 17. The gross lines refer to the central film thicknesses and the thin lines refer to the

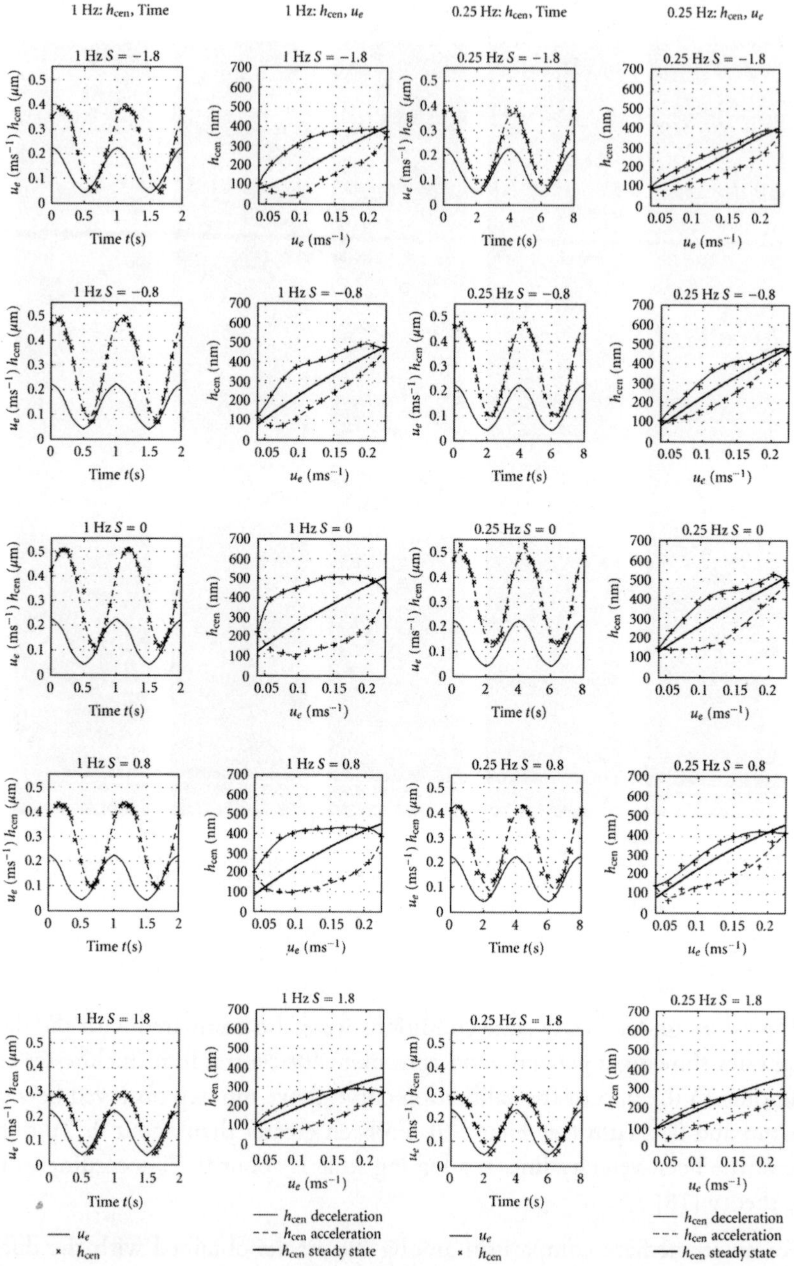

Figure 16: Central film thickness as a function of time and entraining velocity for the two test frequencies 1 Hz and 0. 25 Hz.

minimum film thicknesses, while the solid lines refer to the film thicknesses obtained at test frequency of 1 Hz and the dashed lines refer to the test frequency of 0. 25 Hz. The arrows show the direction of the entraining speed (higher film thicknesses correspond to decelerating speed and lower film thicknesses to accelerating speed).

Some preliminary considerations are reported here, while more detailed explanations of the results are reported in [20].

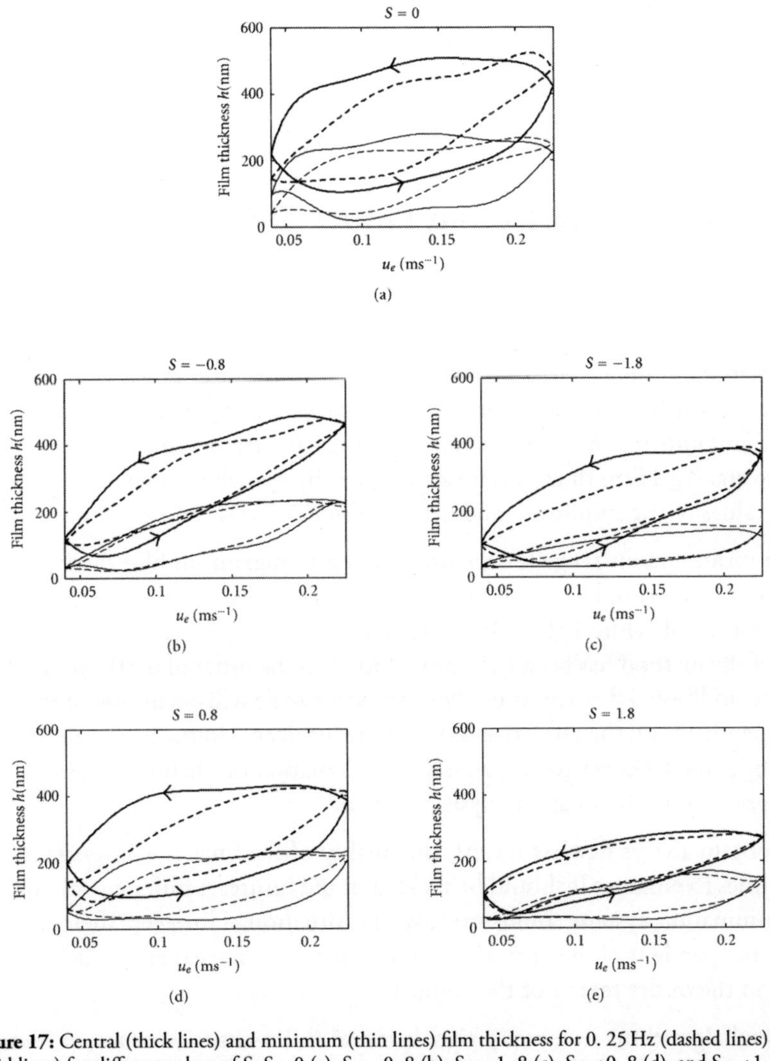

Figure 17: Central (thick lines) and minimum (thin lines) film thickness for 0. 25 Hz (dashed lines) and 1 Hz (solid lines) for different values of S: S = 0 (a), S = −0. 8 (b), S = −1. 8 (c), S = +0. 8 (d), and S = +1. 8 (e).

Due to squeeze effects, the film thickness values are usually higher for decreasing velocity than under steady-state conditions, whereas for increasing velocity they are lower, respectively. Due to the higher accelerations, this effect is greater for the test frequency of 1 Hz than for 0. 25 Hz. At 1 Hz, this hysteresis behaviour is so strong that the most extreme film thicknesses do not coincide with the extremes of the entraining velocity and film thicknesses continue to increase while velocity is already decreasing and reach their maximum only a few measuring points later. When the entraining velocity passes its minimum, the same behaviour can be observed correspondingly. For pure rolling conditions (S = 0), the film thickness loops are bigger than for high slide-to-roll-ratios.

Similarly to the stationary conditions, the highest film thicknesses are reached for pure rolling conditions independently of the test frequency. In addition, also under transient conditions, an asymmetry is found in the film thickness results that indicate lower values of the film thickness for positive S.

Conclusive Remarks and Future Developments

An automated image processing programme has been developed in the Matlab environment in order to determine film thickness maps of EHL point contacts from white light interferograms. It is based on the analysis of the hue channel of interferogram images represented in HSV colour space. It has been particularly developed for processing image series comparatively easily and fast. More insights and information in analysing white light interferograms are gained compared to standard film thickness measurements by eye where often only approximated values of the minimum and central film thickness can be given.

The measurement range is currently limited to maximum film thicknesses of 0. 7 μm and could only be significantly extended by using for example trichromatic instead of white light. On a Hertzian contact, the measurement uncertainty of the method has been determined to be in the order of 0. 01 μm. In EHL contacts, additional deviations of about the same scale will occur due to the pressure dependence on the refractive index of the lubricant. In future developments, this error could be corrected iteratively if a calculation of the pressure profile will be integrated into the image analysis program.

To obtain always flawless results, the quality of the images is important, and possibilities to enhance it should be tackled in the future. A different microscope and illumination system could improve the situation. However, there are also factors independent of the optical set-up that reduces image quality, such as cavitations in the outlet region of the contact.

Several algorithms to unwrap signals found in literature were tested. Finally, a custom solution, which combines a classical integration approach for the phase derivative with a coordinate transformation, was implemented. However, a more sophisticated phase unwrapping algorithm would be necessary to increase the chance of analysing also corrupted image regions successfully.

Similar solutions present in the literature might offer more accurate results, but they are not particularly focused on fast processing of many interferograms as the present programme.

The programme has been used for analysing several images recorded in tests carried out under steady-state as well as transient conditions for different slide-to-roll ratios S and different test frequencies.

The results evidenced a reduction of the film thickness when increasing the percentage of sliding compared to pure rolling conditions. This reduction occurs in an asymmetric way, both for steady-state and transient conditions, showing lower values of the central film thickness for positive S (ball running faster than the disc) than for negative S. This can be related to thermal effects such as the temperature viscosity wedge action. The temperature on the ball surface is lower in the case where the ball runs slower than the disc. For a glass-disc-on-ball configuration, thermal viscosity wedge effects usually occur under conditions of faster disc, for example, the occurrence of Kaneta's dimples [19].

High slide-to-roll ratios values seem to reduce the squeeze effect on the film thickness under transient conditions, particularly for higher test frequencies. The reduction of the amplitude of the film thickness loops is significantly higher for positive Sthan for negative S. This could be related again to thermal effects as the temperature viscosity wedge action. A detailed discussion on these aspects is reported in [20].

References

1. A. Cameron and R. Gohar, "Theoretical and experimental studies of the oil film in lubricated point contact, " Proceedings of the Royal Society of London A, vol. 291, no. 1427, pp. 520–536, 1966.
2. C. A. Foord, W. C. Hammann, and A. Cameron, "Evaluation of lubricants using optical elastohydrodynamics, " Tribology Transactions, vol. 11, no. 1, pp. 31–43, 1968.
3. M. Smeeth and H. A. Spikes, "The influence of slide/roll ratio on the film thickness of an EHD contact operating within the mixed lubrication regime, " in Proceedings of the of 23rd Leeds-Lyon Symposium on Tribology, pp. 695–703, Elsevier, Leeds, UK, September 1996.
4. J. Sugimura, W. R. Jones Jr. , and H. A. Spikes, "EHD film thickness in non-steady state contacts, " Journal of Tribology, vol. 120, no. 3, pp. 442–452, 1998.

5. R. Bassani, E. Ciulli, K. Stadler, and M. Carli, "Lubricated non-conformal contacts under steady state and transient conditions, " in Proceedings of the 17th AIMETA General Conference, Florence, Italy, September 2005.
6. R. Bassani, E. Ciulli, M. Carli, and K. Stadler, "Experimental investigation of transient and thermal effects on lubricated non-conformal contacts, " in Proceedings of the 5th International Conference on Tribology (AITC-AIT '06), Parma, Italy, September 2006.
7. R. Bassani and E. Ciulli, "Lubricant film thickness and shape using interferometry and image processing, " in Elastohydrodynamics-'96: Fundamentals and Applications in Lubrication and Traction: Proceedings of the 23rd Leeds-Lyon Symposium on Tribology (Leeds, UK), pp. 81–90, Elsevier, Amsterdam, The Netherlands, 1997.
8. R. Bassani, E. Ciulli, and R. Squarcini, "A digital image processing method for geometry modeling of a EHL contact, " in Proceedings of the 16th AIMETA Congress of Theoretical and Applied Mechanics, Ferrara, Italy, September 2003.
9. J. Molimard, M. Querry, and P. Vergne, et al. , "Differential colorimetry: a tool for the analysis of fluid film lubrication, " Mécanique & Industries, vol. 3, no. 6, pp. 571–581, 2002.
10. O. Marklund, Interferometric measurements and analysis with applications in elastohydrodynamic experiments, Ph. D. thesis, Luleå University of Technology, Luleå, Sweden, 1999.
11. M. Eguchi and T. Yamamoto, "Film thickness measurement using white light spacer interferometry and its calibration by HSV color space, " in Proceedings of the 5th International Conference on Tribology (AITC-AIT '06), Parma, Italy, September 2006.
12. O. Marklund and L. Gustafsson, "Interferometry-based measurements of oil-film thickness, " Proceedings of the Institution of Mechanical Engineers, Part J: Journal of Engineering Tribology, vol. 215, no. 3, pp. 243–259, 2001.
13. R. S. Berns, Billmeyer and Saltzman's Principles of Color Technology, John Wiley & Sons, New York, NY, USA, 2000.
14. D. C. Ghiglia and L. A. Romero, "Robust two-dimensional weighted and unweighted phase unwrapping that uses fast transforms and iterative methods, " Journal of the Optical Society of America A, vol. 11, no. 1, pp. 107–117, 1994.

15. J. Tribolet, "A new phase unwrapping algorithm, " IEEE Transactions on Acoustics, Speech, and Signal Processing, vol. 25, no. 2, pp. 170–177, 1977.

16. A. K. Jain, Fundamentals of Digital Image Processing, Prentice-Hall, Upper Saddle River, NJ, USA, 1989.

17. O. Marklund and L. Gustafsson, "Correction for pressure dependence of the refractive index in measurements of lubricant film thickness with image analysis, " Proceedings of the Institution of Mechanical Engineers, Part J: Journal of Engineering Tribology, vol. 213, no. 2, pp. 109–126, 1999.

18. B. J. Hamrock, Fundamentals of Fluid Film Lubrication, Mcgraw-Hill, New York, NY, USA, 1994.

19. M. Kaneta and P. Yang, "Effects of thermal conductivity of contacting surfaces on point EHL contacts, " Journal of Tribology, vol. 125, no. 4, pp. 731–738, 2003. 20. E. Ciulli, K. Stadler, and T. Draexl, "The influence of the slide-to-roll ratio on the friction coefficient and film thickness of EHD point contacts under steady state and transient conditions, " in Proceedings of the European Conference on Tribology and Final Conference of COST 532 Action (ECOTRIB '07), vol. 1, pp. 389–400, Ljubljana, Slovenia, June 2007.

Performance of an Orifice Compensated Two-Lobe Hole-Entry Hybrid Journal Bearing

J. Sharana Basavaraja, Satish C. Sharma, and S. C. Jain

ABSTRACT

The work presented in this paper aims to study the performance of a two-lobe hole-entry hybrid journal bearing system compensated by orifice restrictors. The Reynolds equation governing the flow of lubricant in the clearance space between the journal and bearing together with the equation of flow through an orifice restrictor has been solved using FEM and Galerkin's method. The bearing performance characteristics results have been simulated for an orifice compensated nonrecessed two-lobe hole-entry hybrid journal bearing symmetric configuration for the various values of offset factor (δ), restrictor design parameter ($C^{-}S2$), and the value of external load ($W^{-}0$). Further, a comparative study of the performance of a two-lobe hole-entry hybrid journal bearing system with a circular hole-entry symmetric hybrid journal bearing system has also been carried out so that a designer has a better flexibility in choosing a suitable bearing configuration. The simulated numerical results indicate that for the two-lobe symmetric hole-entry hybrid journal bearing system with an offset factor (δ) greater than one provides 30 to 50 percent *larger values of direct stiffness and direct damping coefficients as compared to a circular symmetric hole-entry hybrid journal bearing system.*

Introduction

Hybrid journal bearings are extensively used in machine tools under conditions of heavy loads and higher speeds. The stability and unsteady behavior of the hybrid journal bearings is greatly influenced by bearing geometry, and accordingly various designs have been used by designers to achieve the desired objective. To ensure better dynamic stability than a circular plain journal bearing, multilobe journal bearings are used. Multilobe hydrodynamic journal bearings have been investigated for their antiwhril characteristics by many researchers [1–5]. Pinkus [1] studied the performance of an elliptical journal bearing in terms of steady-state load carrying capacity and power loss using the FDM. Malik [2] theoretically studied an elliptical hydrodynamic journal bearing and compared its performance over a wide range of load conditions and provided the comprehensive design data including the static and dynamic characteristics for the two-lobed journal bearing for different aspect ratios. The study dealing with the effects of surface ellipticity on the dynamically loaded cylindrical bearing was carried out by Goenka and Booker [3] for an optimum bearing shape on the basis of maximizing the minimum film thickness. The notable observation about these studies is that they are all concerned with hydrodynamic journal bearing systems. Few studies dealing with the noncircular multirecess hydrostatic/hybrid journal bearing systems have also been reported in literature recently [6–8]. Ghosh and Satish [7], using a small amplitude perturbation analysis, determined the rotordynamic coefficients of a multilobe hybrid journal bearing system having large recesses. They extended their work further [8] to study the stability of a rigid rotor supported on a multilobe hybrid journal bearing system with short sills following a linear vibration theory for small amplitude oscillations of the journal center about its steady-state position. It was observed that a multilobe recessed hybrid journal bearing system with an offset factor more than one exhibits better dynamic performance than the circular hybrid journal bearings. The recessed journal bearings are unable to generate a substantial hydrodynamic action because recess constitutes a large bearing area, thus leaving very less area for lands. Thus, recessed bearings when operating in hybrid mode at higher speeds are not suitable for heavily loaded applications. Hence, nonrecessed journal bearings, that is, hole-entry hybrid journal bearings were developed to gain the advantage of available large land area to generate substantial hydrodynamic action. Such bearings give better performance than the recessed bearings along with the ease of manufacture and reduced cost of machining [9–12].

Recent developments towards the use of hybrid bearings in high-speed turbopumps and advanced machine tools have further necessitated to focus research activities in the area of hybrid bearings. Most of the available nonrecessed hydrostatic/hybrid bearing studies are generally confined to circular bearings [13–15]. However, a thorough review of literature reveals that the performance of nonrecessed multilobe hole-entry hybrid journal bearing has not yet been investigated. Therefore, a study of a nonrecessed multilobe hole-entry hybrid journal bearing is planned to bridge the gap in literature. In this paper, nonrecessed orifice compensated two-lobe hole-entry hybrid journal bearing is investigated for its static and dynamic characteristics. The performance characteristics of two-lobe hole-entry hybrid journal bearing have been compared with that of a circular hole-entry orifice compensated hybrid journal bearing for the same bearing operating and geometric parameters. Results are presented for various values of offset factor (δ), restrictor design parameter ($C^{-}S2$), and for the value of external load for symmetric hole configuration. The computed results presented in this work are expected to be quite useful to the bearing designers as well as for the academic community.

Analysis

The Reynolds equation governing the laminar flow of an incompressible lubricant in the clearance space of journal and bearing in nondimensional form is expressed as [16, 17]

$$\frac{\partial}{\partial \alpha}\left(\frac{\bar{h}^3}{6}\frac{\partial \bar{p}}{\partial \alpha}\right)+\frac{\partial}{\partial \beta}\left(\frac{\bar{h}^3}{6}\frac{\partial \bar{p}}{\partial \beta}\right)=\Omega\frac{\partial \bar{h}}{\partial \alpha}+2\frac{\partial \bar{h}}{\partial \bar{t}}. \quad . (1)$$

Fluid Film Thickness

The geometry of the symmetric hole-entry circular as well as multilobe hybrid journal bearing has been shown schematically in Figures 1(a) and 1(b). The expression for nondimensional fluid film thickness for a multilobe rigid journal bearing with reference to fixed coordinate axis is given as [6]

$$\bar{h}=\frac{1}{\delta}-(\bar{X}_j+\bar{x}-\bar{X}_L^i)\cos\,\alpha-(\bar{Z}_j+\bar{z}-\bar{Z}_L^i)\sin\alpha, \quad , (2)$$

where $\bar{X}_j$ and $\bar{Z}_j$ are the equilibrium coordinates of the journal center, and $\bar{x}$ and $\bar{z}$ are time dependent perturbation coordinates of the journal center measured from their equilibrium position ($\overline{X}$, $\overline{Z}$ =0 for steady case). $\overline{X}_L^i$ and $\overline{Z}_L^i$ are the lobe center coordinates of ith lobe.

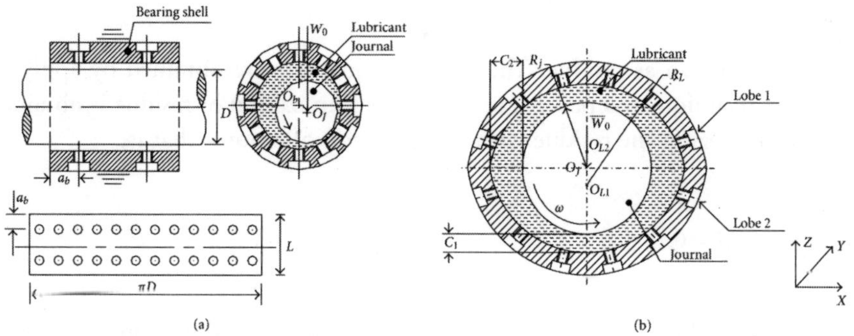

Figure 1: (a) Hole-entry type journal bearing system (symmetric configuration). (b) Two-lobe hole-entry type journal bearing system (symmetric configuration).

Restrictor Flow Equation

The flow rate of the lubricant through the orifice restrictor is defined as [17]

$$\overline{Q}_R = \overline{C}_{S2}(1 - \overline{p}_c)^{1/2}, (3)$$

where the term $\overline{p}_c$ represents the pressure at the hole.

Finite Element Formulation

The lubricant flow field in the clearance space of a circular bearing has been discretized into four-noded isoparametric elements and using the Lagrangian interpolation function, the pressure at a point in the element is bilinearly distributed and expressed approximately as [18]

$$\overline{p} = \sum_{j=1}^{4} N_j \overline{p}_j,, (4)$$

where Nj is elemental shape function, and using the approximate value of $\overline{p}$ (1) can be expressed as

$$\frac{\partial}{\partial\alpha}\left[\frac{\overline{h}^3}{6}\frac{\partial}{\partial\alpha}\left(\sum_{j=1}^{4}N_j\overline{p}_j\right)\right]+\frac{\partial}{\partial\beta}\left[\frac{\overline{h}^3}{6}\frac{\partial}{\partial\beta}\left(\sum_{j=1}^{4}N_j\overline{p}_j\right)\right] \quad (5)$$

where R^e is known as residue. The element equations are obtained by applying Galerkin's technique. As per this technique, minimization of residue is obtained by orthogonalizing the residue with interpolation functions, that is,

$$\iint N_i R^e \, d\alpha\, d\beta = 0. \quad . (6)$$

By integrating the second-order term by parts to obtain C0 continuity and differentiating (4) with respect to $\overline{t}$, the resulting equation for a typical element is obtained in matrix form as follows:

$$[\overline{F}]_{n\times n}\{\overline{p}\}_{n\times 1} = \{\overline{Q}\}_{n\times 1}+\Omega\{\overline{R}_H\}_{n\times 1}+\dot{\overline{x}}_j\{\overline{R}_{x_j}\}_{n\times 1}+\dot{\overline{z}}_j\{\overline{R}_{z_j}\}_{n\times 1}, \quad , (7)$$

where n = total number of nodes in lubricant flow field.

After modification for continuity of flow between bearing and restrictor and incorporating appropriate boundary condition, the system (7) is solved for nodal pressure and nodal flows.

For an eth element, the elements of the above matrices are defined as follows:

$$\overline{F}_{ij}^e = \iint_{A^e} \overline{h}^3\overline{F}_2\left[\frac{\partial N_i}{\partial\alpha}\frac{\partial N_j}{\partial\alpha}+\frac{\partial N_i}{\partial\beta}\frac{\partial N_j}{\partial\beta}\right]d\alpha\, d\beta,$$

$$\overline{Q}_i^e = \int_{\Gamma^e}\left\{\left(\overline{h}^3\overline{F}_2\frac{\partial\overline{p}}{\partial\alpha}-\Omega\left(1-\frac{\overline{F}_1}{\overline{F}_0}\right)\overline{h}\right)l+\left(\overline{h}^3\overline{F}_2\frac{\partial\overline{p}}{\partial\beta}\right)m\right\}N_i\, d\overline{\Gamma}^e,$$

$$\overline{R}_{H_i}^e = \iint_{A^e}\left\{\left(1-\frac{\overline{F}_1}{\overline{F}_0}\right)\overline{h}\right\}\frac{\partial N_i}{\partial\alpha}\, d\alpha\, d\beta,$$

$$\overline{R}^{e}_{X_{Ji}} = \iint_{A^e} N_i \cos\alpha \, d\alpha \, d\beta,$$

$$\overline{R}^{e}_{Z_{Ji}} = \iint_{A^e} N_i \sin\alpha \, d\alpha \, d\beta,$$

where l and m are the directions cosines and i, j=1, , ..., n_l^e(number of nodes per element).

Boundary Conditions

The boundary conditions for the lubricant flow field are as follows.

(1) Nodes situated on the external boundary of the bearing have zero pressure,

$$\overline{p}|_{\beta=\mp 1.0} = 0.0. \quad (8)$$

(2) The nodes situated on a hole have equal pressure.

(3) Flow of lubricant through the restrictor is equal to the bearing input flow.

(4) The nodal flows are zero at internal nodes except those situated on holes.

Solution Procedure

The study of an orifice compensated multilobe hole-entry hybrid journal bearing system needs an iterative solution scheme to establish solutions of the flow field system (5) with the restrictor flow (3) as constraint with boundary conditions. Assuming constant viscosity and the steady-state case ($\overline{X}_j$, $\overline{Z}_j = 0$), the lubricant flow field system (5), after adjustment for flow through orifice restrictor (3) and modification for boundary conditions, is solved for a specified journal center position ($\overline{X}_j$, $\overline{Z}_j$) using Gauss elimination technique. If the solution is to be obtained for a specified vertical external load, one additional iterative loop is needed to establish the equilibrium journal center position using the following equations:

$$\overline{F}_x = 0, \qquad \overline{F}_z - \overline{W}_0 = 0. \cdot (9)$$

The fluid-film reaction terms $\overline{F}_x$ and $\overline{F}_z$ in (9) are expanded by Taylor series about the ith journal center position, and the increments ($\Delta \overline{X}^i_j$, $\Delta \overline{Z}^i_j$) on the-journal coordinates are obtained. Iterations are continued until the following convergence criterion is not satisfied:

$$\left[\frac{\left(\left(\Delta\overline{X}_j^i\right)^2 + \left(\Delta\overline{Z}_j^i\right)^2\right)^{1/2}}{\left(\left(\overline{X}_j^i\right)^2 + \left(\overline{Z}_j^i\right)^2\right)^{1/2}}\right] \times 100 < 0.001,, \quad (10)$$

where $\overline{X}_j^i$, $\overline{Z}_j^i$ are the coordinates of the *i*th journal center position. The overall iterative solution scheme is presented in Figure 2.

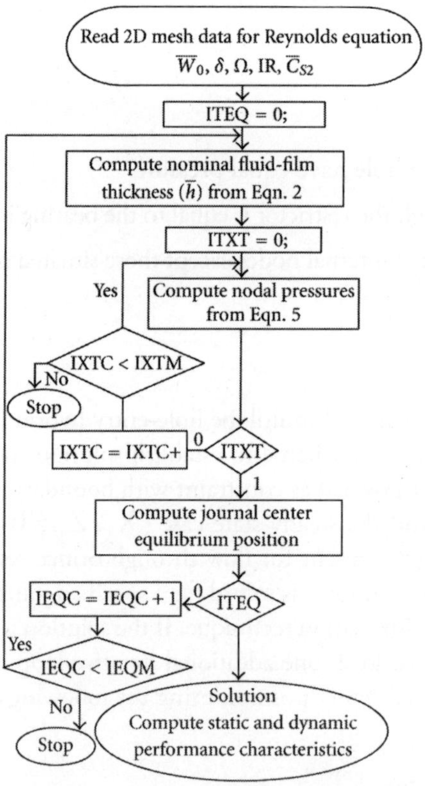

IEQC: Iteration counter for journal center equilibrium
IXTC: Iteration counter for extent of fluid film
IEQM: Maximum number of iteration for journal center equilbrium
IXTM: Maximum number of iteration for extent of fluid film
IR: Code for type of restrictor

Figure 2: Overall iterative scheme.

Fluid Film Stiffness and Damping Coefficients

Fluid-film stiffness and damping coefficients are computed using the expressions given below.

Fluid-film stiffness coefficients:

$$\bar{S}_{ij} = -\frac{\partial \bar{F}_i}{\partial \bar{q}_J}, \ (i = x, z), \ (11)$$

where:

i = direction of force or moment;

$\bar{q}_J$ = direction of journal center displacement ($\bar{q}_J = \bar{X}_J, \bar{Z}_J$).

Fluid-film damping coefficients:

$$\bar{C}_{ij} = -\frac{\partial \bar{F}_i}{\partial \dot{\bar{q}}_J}, \quad (i = x, z), \quad (12)$$

here, $\bar{q}_J$ represents the velocity component of journal center ($\bar{q}_J = \bar{X}_J, \bar{Z}_J$).

Results and Discussion

The performance characteristics for circular and two-lobe symmetric hole-entry hybrid journal bearing have been computed using the solution scheme as discussed earlier. The results have been computed for various values of offset factor (δ) and restrictor design parameter ($\bar{C}_{S2}$). The chosen values of bearing operating parameter are most generally used values as given in Table 1.

Table 1: Bearing operating and geometric parameters.

Bearing aspect ratio (λ)	1
Land width ratio ($\bar{a}_b$)	0.25
Restrictor design parameter ($\bar{C}_{S2}$)	0.02–0.1
Offset factor (δ)	0.5, 0.75, 1.0, 1.25, 1.5
External load ($\overline{W}_0$)	1.25
Number of holes per row	12
Number of rows	2
Concentric pressure ratio (β^*)	0.5
Type of restrictor	orifice

Table 2: Comparison of performance characteristics of two-lobe hydrodynamic journal bearing, L/D = 1, Ω = 1, δ=0. 5. $\overline{W}_0$= 5. 0.

Characteristics	Present	Ref. [19]
ε	0.3426	0.340
ϕ	80.997	82.100
$\overline{S}_{xx}/\overline{W}_0$	1.203	1.110
$\overline{S}_{xz}/\overline{W}_0$	1.812	1.700
$\overline{S}_{zx}/\overline{W}_0$	−3.5004	−3.610
$\overline{S}_{zz}/\overline{W}_0$	5.1544	5.130
$\overline{C}_{xx}/\overline{W}_0$	2.801	2.620
$-(\overline{C}_{xz} = \overline{C}_{zx})/\overline{W}_0$	0.2568	0.238
$\overline{C}_{zz}/\overline{W}_0$	9.3348	9.430

In order to check the validity of the computed results obtained from the developed program, the computed results have been compared with those already available in the literature. To the best of author's knowledge, no results are available in the literature for the performance characteristics of nonrecessed multilobe hole-entry hybrid journal bearing. Hence, to validate the results, first the values of fluid film reaction ($\overline{F}_0$) for orifice compensated hole-entry hydrostatic journal bearing are computed and compared with that of Rowe et al. [10]. The results compare well for wide range of eccentricity ratio (ε) as shown in Figure (3). Further, using Reynolds cavitation boundary condition, the performance characteristics of two-lobe hydrodynamic journal bearing are compared with the data of Lund and Thomsen [19]. The results match very well as shown in Table 2.

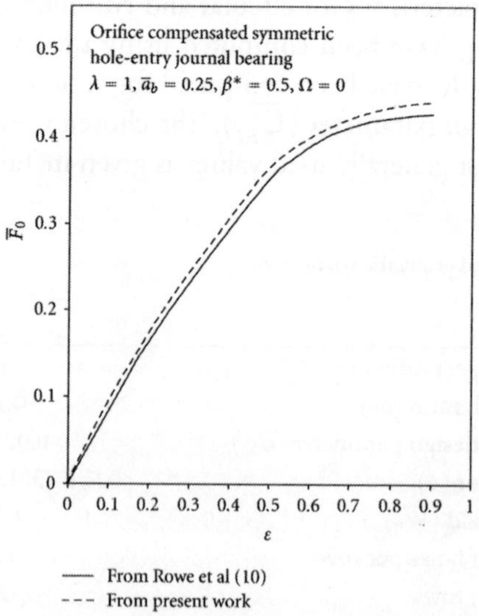

Figure 3: Variation of external load carrying capacity ($\overline{F}_0$) with eccentricity (ε).

The static and dynamic performance characteristics which include maximum pressure ($\overline{p}$), minimum fluid film thickness ($\overline{h}_{min}$), bearing flow ($\overline{Q}$), stiffness coefficients ($\overline{S}_{ij}$), damping coefficients ($\overline{C}_{ij}$), and threshold speed ($\overline{\omega}_{th}$) have been presented for various values of offset factors (δ), restrictor design parameter ($\overline{C}_{S2}$), and external load ($\overline{W}_0$). The results are presented through Figures (4–14) and discussed in the following paragraph.

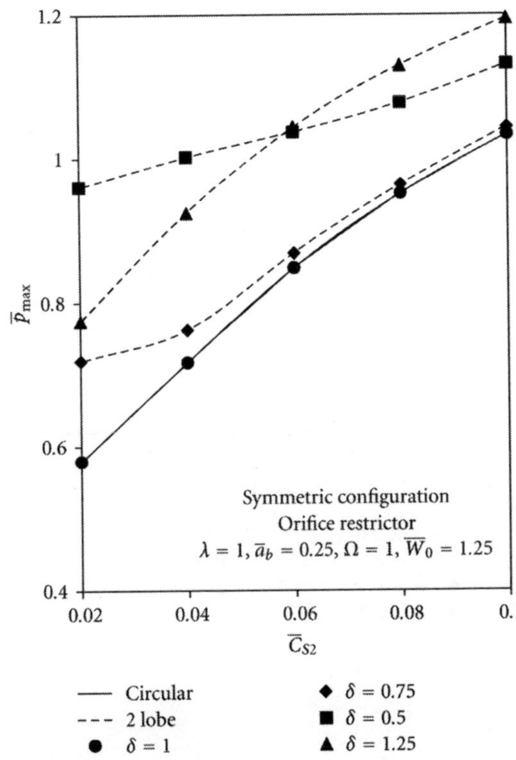

Figure 4: Variation of maximum pressure ($\overline{p}$).

Figure 4 shows the variation of maximum pressure ($\overline{p}$). It has been observed that the value of maximum pressure ($\overline{p}$) for a specified external load increases with restrictor design parameter ($\overline{C}_{S2}$). From the graph, it may be observed for the given value of external load ($\overline{W}_0$) and the restrictor design parameter ($\overline{C}_{S2}$) that the value of maximum pressure is higher for two-lobe hole-entry hybrid journal bearing. It is also observed that the value of maximum pressure ($\overline{p}$) increases in case of two-lobe hole-entry hybrid journal bearing with offset factor (δ) more than one compared to circular hole-entry hybrid journal bearing. The decrease in the value of fluid film thickness ($\bar{h}$) owing to the change of geometry of the bearing is responsible for the increase in the value of maximum pressure ($\bar{p}$).

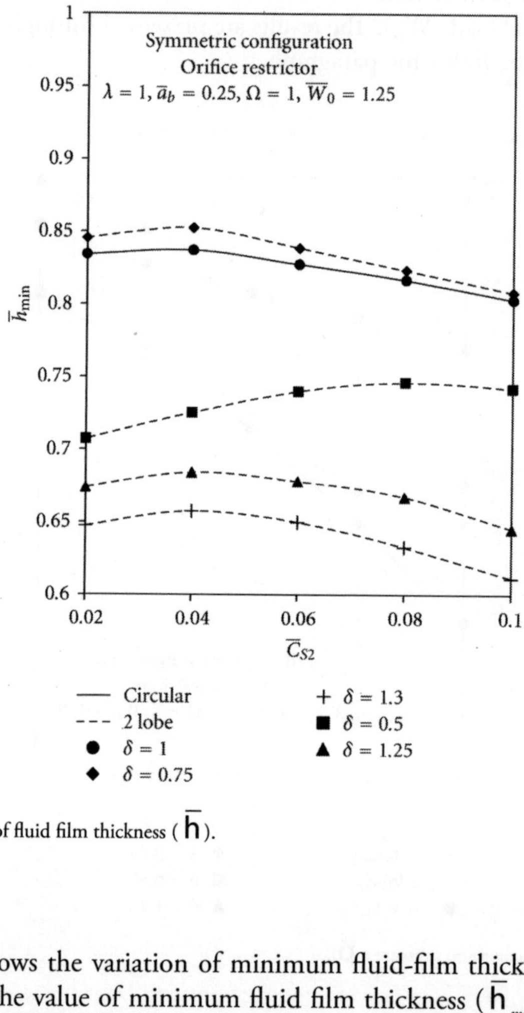

Figure 5: Variation of fluid film thickness ($\bar{h}$).

Figure 5 shows the variation of minimum fluid-film thickness ($\bar{h}_{min}$). It is observed that the value of minimum fluid film thickness ($\bar{h}_{min}$) for a specified external load reduces with an increase in restrictor design parameter ($\bar{C}_{S2}$) for all the cases of offset factor (δ) except for the case of offset factor δ = 0. 5. At a constant external load, the two-lobe hole-entry hybrid journal bearing with offset factor (δ) more than one operates at higher eccentricities compared to circular hole-entry hybrid journal bearing. Also Figure 5 indicates that the two-lobe hole-entry hybrid journal bearing with offset factor δ = 0. 75 operates with higher value of minimum fluid film thickness ($\bar{h}_{min}$) compared to other cases of offset factor (δ). The increase or decrease in the value of minimum fluid film thickness ($\bar{h}_{min}$) has been found of the order of 0. 5 < . 75 >1. 0 >1. 25 >1. 3. The maximum value of minimum fluid film thickness ($\bar{h}_{min}$) is not exactly for δ = 1. 0, but rather it is observed to be at the values between δ = 0. 75 and δ = 0. 9. Similar tend has been reported by Goenka and Booker [3]. This trend may be attributed due to the clearance change in the bearing for the values of offset factor (δ). It may be observed that a desired value of h¯min may be obtained in a two-lobe hole-entry journal bearing by carefully selecting the value of restrictor design parameter ($\bar{C}_{S2}$) and offset factor (δ).

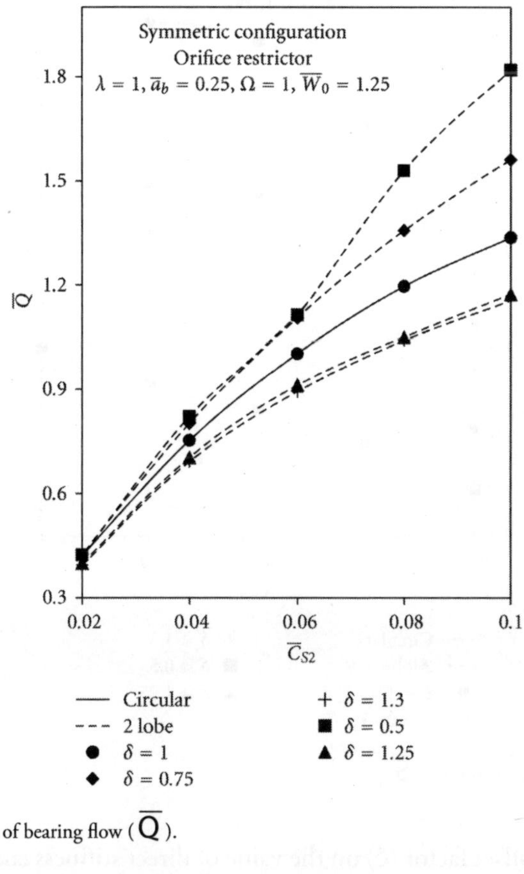

Figure 6: Variation of bearing flow ($\bar{Q}$).

At a constant value of external load ($\overline{W}_0$), the bearing flow ($\bar{Q}$) in Figure 6 is found to be lower for the case of two-lobe hole-entry hybrid journal bearing with offset factor (δ) greater than one compared to bearing flow in circular hole-entry hybrid journal bearing. But the bearing flow is higher for the case of two-lobe hole-entry journal bearing with offset factor less than one. This change in bearing flow may be accounted for the change in profile of the bearing due to change in offset factor (δ). The decrease in the flow rate in case of two-lobe hole-entry hybrid journal bearing with offset greater than one minimizes the pumping power.

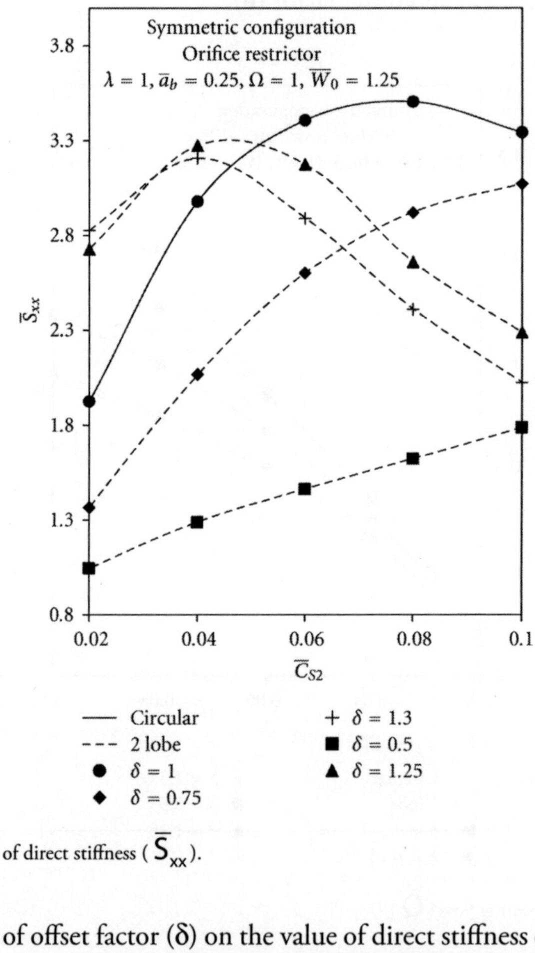

Figure 7: Variation of direct stiffness ($\bar{S}_{xx}$).

The effects of offset factor (δ) on the value of direct stiffness coefficients ($\bar{S}_{xx}$, $\bar{S}_{zz}$) have been presented in Figures 7 and 8. From Figure 7, it may be observed that there exists a particular value of restrictor design parameter ($\bar{C}_{S2}$) at which the value of direct stiffness coefficient ($\bar{S}_{xx}$) is the maximum for a particular case of offset factor (δ). As seen from Figure 7, at a value of restrictor design parameter ($\overline{C}_{S2}$=. 04), the two-lobe hole-entry hybrid journal bearing with an offset factor δ = 1. 25 has the highest value of direct stiffness coefficient ($\overline{S}_{xx}$) and at a value of restrictor design parameter ($\overline{C}_{S2}$=. 08), the hole-entry hybrid journal bearing with an offset factor δ = 1. 0 has the highest value of direct stiffness coefficient ($\overline{S}_{xx}$). The results from Figure 7 indicate that the restrictor design parameter ($\overline{C}_{S2}$) becomes an important parameter when the designer aims to maximize the value of ($\overline{S}_{xx}$) at constant external load ($\overline{W}_0$). Figure 8 shows, at a constant external load $\overline{W}_0$, fluid film stiffness coefficients ($\overline{S}_{zz}$) which in general show an initial increasing trend and then decrease with an increase in restrictor design parameter ($\overline{C}_{S2}$) except for the case of offset factor δ = 0. 5. The two-lobe hole-entry hybrid journal bearing indicates a higher value of direct fluid-film stiffness coefficients ($\overline{S}_{zz}$) compared with circular hole-entry hybrid journal bearing for a value of offset factor (δ) greater than one for a chosen load.

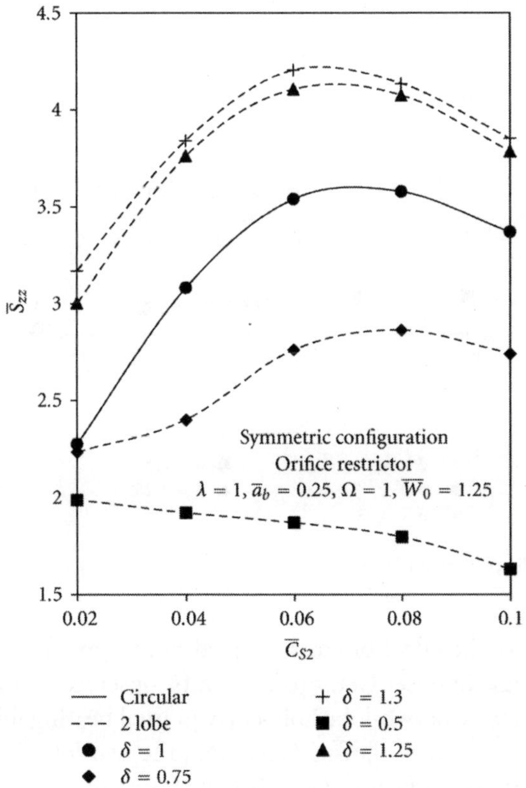

Figure 8: Variation of direct stiffness ($\overline{S}_{zz}$).

Figure 9 shows the effect of direct fluid-film stiffness coefficients ($\overline{S}_{zz}$) with an increase in offset factor (δ). The value of direct fluid-film stiffness coefficient ($\overline{S}_{zz}$) gets decreased with an increase in the value of external load ($\overline{W}_0$) for all the cases of offset factor (δ). Further, it may be observed that the value of direct fluid-film stiffness coefficient ($\overline{S}_{zz}$) is maximum in case of two-lobe hole entry hybrid journal bearing with an offset factor δ = 1. 3.

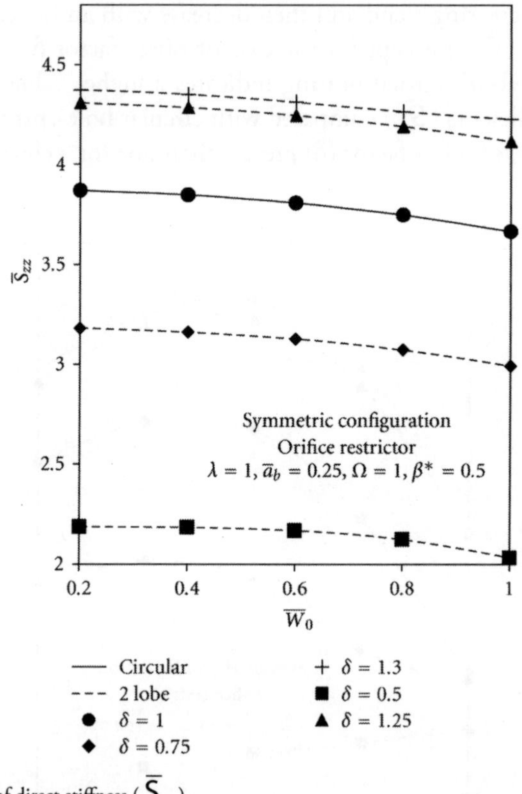

Figure 9: Variation of direct stiffness ($\overline{S}_{zz}$).

Figure 10 shows the effect of cross coupled fluid-film stiffness coefficients ($\overline{S}_{zx}$) with an increase in offset factor (δ). It can be observed from the figure that the values of $\overline{S}_{zx}$ in case of two-lobe hole-entry journal bearing follow the similar trend for the different values of offset factor (δ). The two-lobe hole-entry hybrid journal bearing with offset factor of 0. 5 has the least value of $\overline{S}_{zx}$. The variation of direct damping coefficients ($\overline{C}_{xx}$, $\overline{C}_{zz}$) with offset factor (δ) is shown in Figures 11 and 12. At constant load ($\overline{W}_0$), damping coefficients ($\overline{C}_{xx}$, $\overline{C}_{zz}$)

are increased with an increase of offset factor (δ). It is observed from Figures 11 and 12 that the two-lobe hole-entry hybrid journal bearing with offset factor (δ) greater than one exhibits higher value of direct damping coefficients ($\overline{C}_{xx}$, $\overline{C}_{zz}$) compared to that of circular hole-entry hybrid journal bearing.

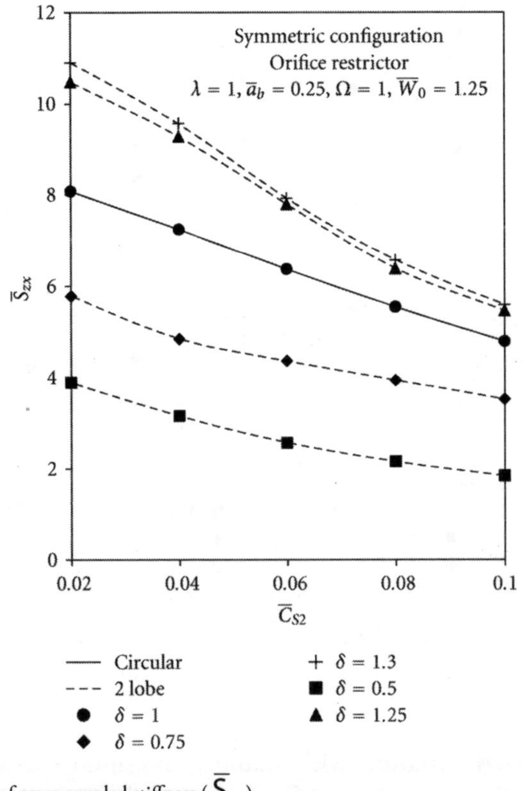

Figure 10: Variation of cross coupled stiffness ($\overline{S}_{zx}$).

Figure 13 shows the variation of direct fluid-film coefficient ($\overline{C}_{zz}$) against restrictor design parameter ($\overline{C}_{S2}$). The value of direct fluid-film damping coefficient ($\overline{C}_{zz}$) gets increased with an increase in offset factor (δ) for the chosen value of external load ($\overline{W}_0$). Further, it may be observed that the value of direct fluid-film damping coefficient ($\overline{C}_{zz}$) decreases with an increase in the value of restrictor design parameter ($\overline{C}_{S2}$) for all the cases of offset factor (δ) except for the case of offset factor δ=0. 5. For the case of offset factor δ=0. 5, the value of direct fluid-film damping coefficient ($\overline{C}_{zz}$) nearly remains a constant with an increase in restrictor design parameter ($\overline{C}_{S2}$).

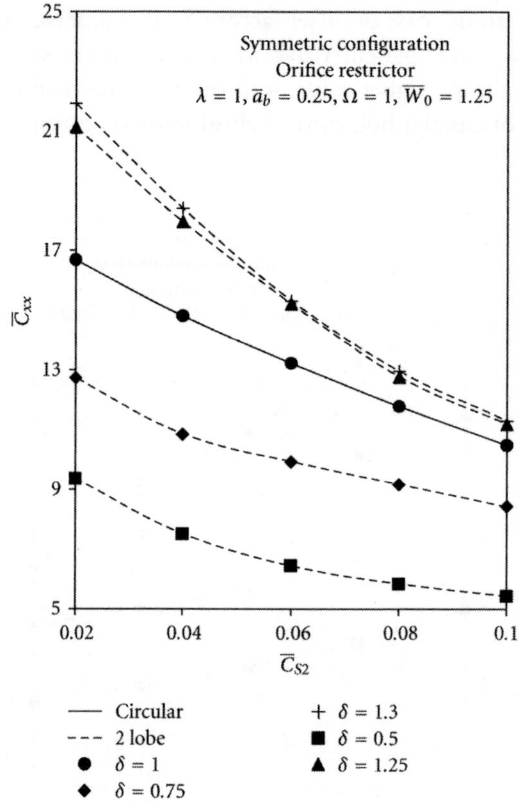

Figure 11: Variation of direct damping ($\overline{C}_{xx}$).

Figure 14 shows variation with stability threshold speed ($\overline{\omega}_{th}$) against the restrictor design parameter ($\overline{C}_{S2}$). For a chosen value of load at a constant speed, the system is asymptotically stable when the operating speed of the journal (ωJ) is less than stability threshold ($\overline{\omega}_{th}$), that is, $\omega_J < \omega_{th}$. From Figure 7, it may be observed that there exists a particular value of restrictor design parameter ($\overline{C}_{S2}$) at which the value of stability threshold speed ($\overline{\omega}_{th}$) is the maximum for a particular case of offset factor (δ). As seen from the Figure 7, at a value of restrictor design parameter ($\overline{C}_{S2}$=. 06), the two-lobe hole-entry hybrid journal bearing with an offset factor δ=1. 25 has the highest value of stability threshold speed ($\overline{\omega}_{th}$) and at a value of restrictor design parameter ($\overline{C}_{S2}$=. 08), the hole-entry hybrid journal bearing with an offset factor δ=1. 0 has the highest value of stability threshold speed ($\bar{\omega}_{th}$). The results from Figure 7 indicate that the restrictor design parameter ($\bar{C}_{S2}$) becomes an important parameter when the designer aims to maximize the value of(ω^-th) at constant external load ($\bar{W}_0$).

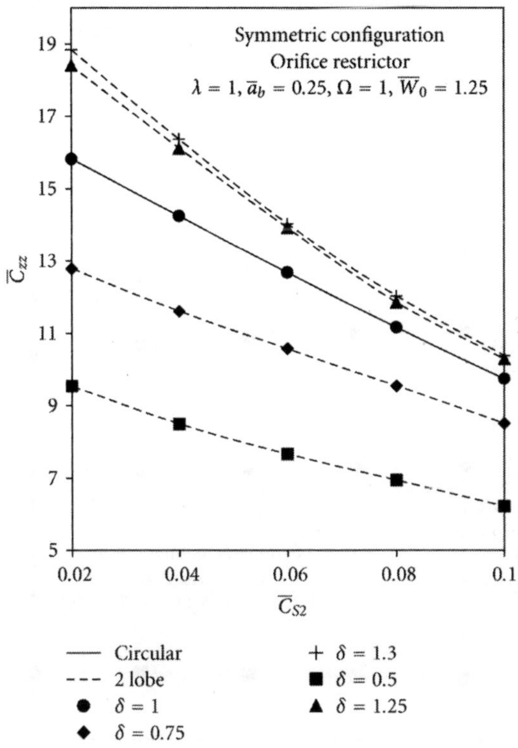

Figure 12: Variation of direct damping ($\bar{C}_{zz}$).

Conclusions

On the basis of the theoretical results presented in the paper, the following conclusions have been drawn.

(1) Two-lobe hole-entry hybrid journal bearing with an offset factor (δ) greater than one requires a reduced quantity of lubricant as compared to circular hybrid hole-entry journal bearing, hence reduced pumping power is

needed. (2) A two-lobe hole-entry hybrid journal bearing system with an offset factor (δ) greater than one provides a larger value of stiffness ($\overline{S}_{zz}$) and damping ($\overline{C}_{xx}$) coefficients. These values are found to be of the order of 15. 9% and 36. 2%, respectively more vis-a-vis a circular hole-entry journal bearing system for the same value of external load ($\overline{W}_0$) and for the same bearing geometric parameters. (3)A two-lobe hole-entry journal bearing system with offset factor (δ) greater than one provides enhanced value of stability threshold speed margin of the order of 14. 4% to that of circular hole-entry hybrid journal bearing system.

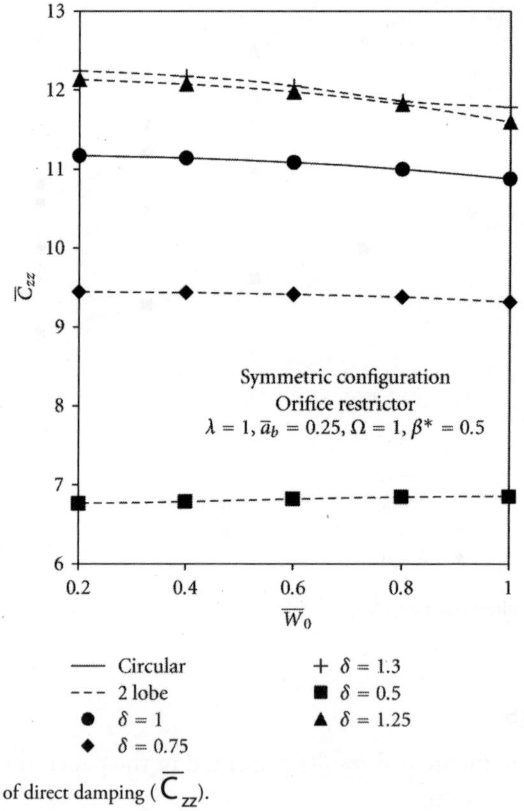

Figure 13: Variation of direct damping ($\overline{C}_{zz}$).

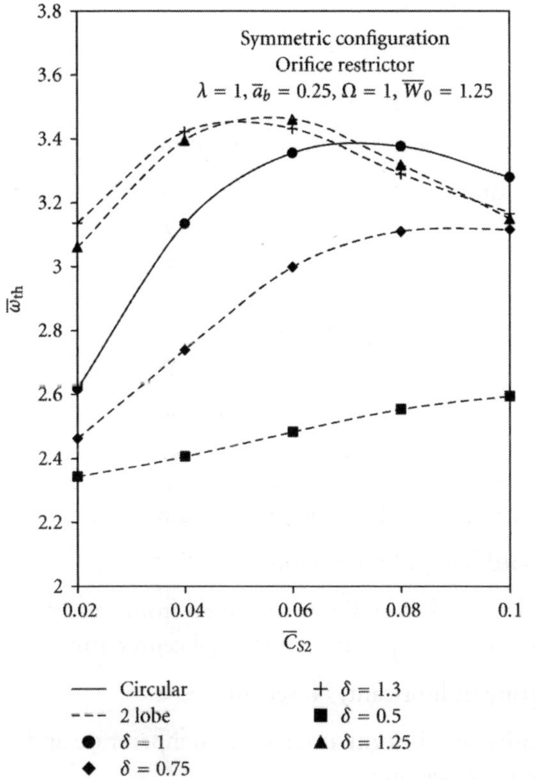

Figure 14: Variation of threshold speed ($\overline{\omega}_{th}$).

Nomenclature

a_b:	Bearing land width, mm
c:	Radial clearance, mm
C_1=c, C_2:	Clearance due to circumscribed, inscribed circle on the bearing, mm
C_{ij}:	Damping coefficients (*i, j=x, z*), N. sec. mm^{-1}
d_0:	Orifice diameter, mm
e:	Journal eccentricity, mm
h, L:	Fluid-film thickness, bearing length mm
t:	Time, second

F: Fluid film reaction ($\partial h/\partial t \neq 0$), N

F_0: Fluid film reaction ($\partial h/\partial t=0$), N

g: Acceleration due to gravity, m. sec^{-2}

R_J, R_b: Radius of journal and bearing, respectively, mm

R_L: Radius of lobe, mm

p, p*: Pressure, concentric design pressure N. mm^{-2}

Q: Bearing flow, mm3. sec^{-1}

S_{ij}: Stiffness coefficients, (*i, j=x, z*), N. mm^{-1}

W_0: External load, N

X, Y, Z: Cartesian coordinates

X_J, Z_J: Coordinates of steady-state equilibrium journal center from geometric center of bearing, coordinates, mm

X_L^i, Z_L^i : Lobe center coordinates of the ith lobe

x, z: Horizontal and vertical co-ordinate measured from steady-state equilibrium position of journal center, mm

μ: Dynamic viscosity of lubricant, N. sec. m^{-2}

μ_r: Dynamic viscosity of lubricant at reference temperature and pressure, N. sec. m^{-2}

ω_J, ω_{th}: Journal rotational speed, threshold speed rad. sec−1

ω_I: (g/c)1/2, rad. sec−1

ϕ: Attitude angle, rad

ρ: Density of lubricant, kg. mm−3

ψ_d: Coefficient of discharge for orifice

Nondimensional Parameters

$\bar{a}_b = a_b/L$: Land width ratio

$\bar{C}_{ij} = C_{ij}(c^3/\mu R_J^4)$

$\bar{C}_{S2} = 1/12(3\pi d_0^2 \mu_r \psi_d / c^3)(2/\rho p_s)^{1/2}$: Orifice design parameter

$(\bar{F}, \bar{F}_0) = (F, F_0)/p_s R_J^2$

$(\bar{h}) = (h)/c$

$(\bar{p}, \bar{p}_s, \bar{p}_c, \bar{p}_{max}) = (p, p_s, p_c, p_{max})/p_s$

$\bar{Q}=Q(\mu/c^3p_s)$, $\bar{S}_{ij}=S_{ij}(c/p_s R_J^2)$

$\bar{W}_0=W_0/p_s R_J^2$

$(\bar{X}_J, \bar{x}, \bar{Z}_J, \bar{z})=(X_J, x, Z_J, z)/c$

$(\bar{X}_L^i, \bar{Z}_L^i)=(X_L^i, Z_L^i)/c$

$(\alpha, \beta)=(X, Y)/R_J$: Circumferential axial coordinates

$\beta* = $ Concentric design pressure ratio, p^*/p_s

$\varepsilon = e/c$: Eccentricity ratio

$\lambda=L/D$: Aspect ratio

$\Omega=\omega_J(\mu R_J^2/c^2p_s)$: Speed parameter

$\bar{\omega}_{th}=\omega_{th}/\omega_I$

$\bar{t}=t(c^2p_s/\mu R_J^2)$

$\delta=C_1/C_2$: Offset factor

O_J: Journal center

O_{Li}: Lobe center

Matrices

$[\bar{F}]$:	Fluidity matrix
$\{\bar{p}\}$:	Nodal pressure vector
$\{\bar{Q}\}$:	Nodal flow vector
$\{\bar{R}_H\}$:	Vector due to hydrodynamic terms
$\{\bar{R}_{Xj}\}, \{\bar{R}_{Zj}\}$:	Right-hand side vectors due to journal center velocities

Subscripts and Superscripts

0:	Steady-state solution
J:	Journal
R:	Restrictor
s:	Supply
–:	Corresponding nondimensional parameter
l:	Lobe
i:	Lobe number

References

1. O. Pinkus, "Analysis of elliptical bearing, " Transactions of the ASME, vol. 78, pp. 965–973, 1956.
2. M. Malik, "A comparative study of some two-lobed journal bearing configurations, " Tribology Transactions, vol. 26, no. 1, pp. 118–124, 1983.
3. P. K. Goenka and J. F. Booker, "Effect of surface ellipticity on dynamically loaded cylindrical bearing, " Transactions of the ASME, Journal of Lubrication Technology, vol. 105, pp. 1–12, 1983.
4. Y. -M. Yan, Q. -H. Yan, and Q. An, "An numerical study on the influences of preload factor on the properties of three-lobe journal bearing, " Journal of East China University of Science and Technology, vol. 32, no. 5, pp. 612–616, 2006.
5. R. Pai and B. C. Majumdar, "Stability of submerged four-lobe oil journal bearings under dynamic load, " Wear, vol. 154, no. 1, pp. 95–108, 1992.
6. M. K. Ghosh and A. Nagraj, "Rotordynamic characteristics of a multilobe hybrid journal bearing in turbulent lubrication, " Proceedings of the Institution of Mechanical Engineers, Part J: Journal of Engineering Tribology, vol. 218, no. 1, pp. 61–67, 2004.
7. M. K. Ghosh and M. R. Satish, "Rotordynamic characteristics of multilobe hybrid bearings with short sills—part I, " Tribology International, vol. 36, no. 8, pp. 625–632, 2003.
8. M. K. Ghosh and M. R. Satish, "Stability of multilobe hybrid bearing with short sills—part II, " Tribology International, vol. 36, no. 8, pp. 633–636, 2003.
9. D. Khosal and W. B. Rowe. , "Fluid film journal bearings operating in a hybrid mode—part 2: experimental investigations, " Transactions of the ASME, Journal of Lubrication Technology, vol. 103, no. 4, pp. 566–573, 1981.
10. W. B. Rowe, S. X. Xu, F. S. Chong, and W. Weston, "Hybrid journal bearings with particular reference to hole-entry configurations, " Tribology International, vol. 15, no. 6, pp. 339–348, 1982.
11. S. Yoshimoto, W. B. Rowe, and D. Ives, "A theoretical investigation of the effect of inlet pocket size on the performance of hole-entry hybrid journal bearings employing capillary restrictors, " Wear, vol. 127, no. 3, pp. 307–318, 1988.

12. W. B. Rowe, D. Koshal, and K. J. Stout, "Slot-entry bearings for hybrid hydrodynamic and hydrostatic operation, " Journal of Mechanical Engineering Science, vol. 18, no. 2, pp. 73–78, 1976.

13. L. Guo and J. Zhu, "Analysis of circular slot entry hybrid bearing—part I, " Journal of Hunan University Natural Sciences, vol. 23, no. 3, p. 103, 1996.

14. J. C. T. Su and K. N. Lie, "Rotation effects on hybrid hydrostatic/ hydrodynamic journal bearings, " Industrial Lubrication and Tribology, vol. 53, no. 6, pp. 261–267, 2001.

15. H. C. Garg, H. B. Sharda, and V. J. Kumar, "On the design and development of hybrid journal bearings: a review, " Tribotest, vol. 12, no. 1, pp. 1–19, 2006.

16. S. C. Sharma, R. Sinhasna, and S. C. Jain, "An elastohydrostatic study of hole-entry hybrid flexible journal bearings with capillary restrictors, " Tribology International, vol. 26, no. 2, pp. 93–107, 1993.

17. S. C. Sharma, S. C. Jain, and N. M. M. Reddy, "A study of non-recessed hybrid flexible journal bearing with different restrictors, " Tribology Transactions, vol. 44, no. 2, pp. 310–317, 2001.

18. K. H. Huebner, The Finite Element Method for Engineers, A Wiley-Interscience Publication, John Wiley & Sons, New York, NY, USA, 1975.

19. J. W. Lund and K. K. Thomsen, "A calculation method and data for the dynamic coefficients of oil lubricated journal bearings, " in Topics In Fluid Film Bearing and Rotor Bearing System Design and Optimization, pp. 1–28, ASME, New York, NY, USA, 1978.

Experimental Investigation into the Tractive Prerolling Behavior of Balls in V-Grooved Tracks

Kris De Moerlooze and Farid Al-Bender

ABSTRACT

In a rolling element system, the period of transition between motion commencement and the attainment of steady state, gross rolling, and termed prerolling is of common concern to many engineering applications. This region is marked by hysteresis friction behavior, with a characteristic friction-displacement curve, which is in particular relevant to motion characterization and control issues. In a previous paper, the authors carried out a theoretical analysis of tractive prerolling, leading to a model for simulating this phenomenon. The present paper is dedicated to the experimental investigation of tractive prerolling friction behavior, including validation of the theoretical model. Firstly, a kinematic analysis of the rolling motion in V-grooved tracks is carried out. Secondly, the influence of the normal load on the frictional behavior, in prerolling up to the attainment of gross rolling, is investigated on a dedicated test setup. Finally, the newly developed theoretical model is validated by comparison with the experimental results. Satisfactory agreement is obtained between theory and experiment.

Introduction

The characterization of rolling friction is a pertinent issue in many fields ranging from rolling stock, aircraft servo systems, and rotary and linear positioning systems to the control of miniature reading/writing heads, friction drives, and high precision systems. This topic has therefore been gaining importance over the last decades, owing to the increasing demands of industrial applications on positioning accuracy, wear, and controllability. A common concern in all these applications is the rolling element system's behavior between motion reversal points, in other words, the period between motion commencement and the attainment of gross rolling. This period is referred to as prerolling, which is akin to presliding in sliding contacts [1–3]. Prerolling is marked by a clear hysteresis behavior with nonlocal memory, which lies at the root of the strongly nonlinear dynamic behavior that characterizes systems comprising rolling elements [4, 5]. The design parameters of rolling element systems, such as preload, ball diameter, and track groove angle, evidently, have their influence on the frictional behavior of the system. Despite this fact, a coherent, systematic theory that correlates behavior with parameters has not yet received the attention it deserves in the literature, owing to many reasons, in particular the complexity and the strong nonlinearity of the rolling contact problem.

Rolling resistance or, as it is often called, rolling friction arises from two basic mechanisms: internal mechanical losses in the bulk of rolling objects and external tractions in the rolling contact patch. In case of elastic rolling objects (such as metallic objects), the former is generally negligible in magnitude in comparison with the latter. This paper is concerned with tractive rolling; more specifically that, which is caused by creepages in the rolling motion, in particular, spin creepage.

When an elastic body of revolution rolls tractively over another, the traction field in the contact patch changes progressively with the distance traversed from its initial distribution until it reaches a certain steady-state distribution. This distribution, which is independent of the initial field prior to commencement of rolling, does not vary with further rolling. This eventual rolling regime is termed gross rolling; the period building up to it, from commencement of rolling, is termed the prerolling regime. The resultant traction in this regime is characterized by rate-independent hysteresis behavior with nonlocal memory in function of the traversed displacement [2]. In this reference, an extensive experimental study of prerolling behavior of a ball in V-groove application has been reported, where the reader may also find a short overview of literature around the subject. The conclusions and correlations presented in that study remained, however, empirical.

Recently, the present authors have developed a theoretical model, which is able to simulate tractive prerolling (and transition to gross rolling) in an arbitrary rolling contact [6]. The present paper introduces on the one hand an experimental validation of that model and on the other hand a check of the validity of Amontons' law as applied to rolling contacts.

Bearings can be considered as one of the most important machine elements in regard to accuracy of motion. However, only few types of bearings, notably cylindrical and tapered roller bearings, contain rolling elements which undergo pure rolling, that is, rolling without creepage, which leads to microslip. In other types, such as pivot bearings, axially loaded deep groove ball bearings, angular contact bearings, and recirculating rolling element guideways, the rolling motion of the rolling elements comprises a significant amount of microslip which in those cases is responsible for the greatest part of the frictional losses. This microslip depends on the relative velocity (or creepage) between the contacting surfaces and on the elasticity.

Other applications where rolling with creepage appears are gears and friction wheel transmissions. The study of rolling contacts comprising microslip becomes therefore an important research issue.

Compared to steady state rolling friction, only a few experimental studies in the field of prerolling have been conducted. Liu et al. [7] perform some experiments on a newly developed test apparatus in this field. They distinguish three regimes for the coefficient of rolling friction, namely, a gradually increasing coefficient in the elastic deformation regime, a peak in the transition zone, and a constant value in the gross rolling regime. Additionally, the wear behavior of steel is investigated under different conditions. The breakaway rolling friction (i. e. , the peak value prior to commencement of gross rolling) is studied by Budinski [8] in a case study in which he presents a modified test procedure for the breakaway coefficient of rolling friction for rolling element bearings.

Endo et al. [9] study the nonlinear spring characteristic of a rolling element in the transient zone. They link the energy losses in the contact to the overall damping characteristics of the rolling system and confirm experimentally that these losses vary when rolling elements with different diameters are used. Otsuka and Masuda [10] call the prerolling region the "nonlinear spring area" and try to explain the influence of the nonlinear spring behavior (NSB) on the frequency response in some practical case studies.

Besides dry rolling contacts, lubricated contacts have also been investigated in the past. Wang et al. [11] investigate the behavior of EHL films under pure rolling, short stroke reciprocating motion. High-speed color camera measurements are compared to simulations with multigrid techniques.

Xiao et al. [12] have suggested a friction model for dry and lubricated rolling of rough bodies. They observe a linear increasing coefficient of rolling friction with increasing mean contact pressure, up to a maximum limit above which the coefficient of friction stays constant.

Load variations in mechanical systems comprising rolling elements often occur, while their influence on the frictional behavior is neglected in control models. This may impose a limit on the attainable positioning accuracy. Since this issue is of importance in machine design and control, this study deals with the influence of the normal load on the prerolling behavior. The first objective is to verify the effect of the normal load on the overall shape and behavior of the hysteresis loop on an existing test setup. The second objective is the verification of the theory presented in [6] (which is the theoretical part of this study) for the case of rolling with spin creepage.

The paper is organized as follows. Section 2 considers the kinematics of the rolling element system which is used during the experiments and estimates the relative order of magnitude of the various types of possible friction losses. Section 3 gives an overview of the experiments. In Section 4, the hysteresis identification is discussed. A model equation is proposed which is used to fit and process the data for further analysis. In Section 5, the influence of the normal load on the parameters of the model equation is studied. All these aspects provide the prerequisites for the validation of the theoretical model, presented in [6], in Section 6. Finally, appropriate conclusions are drawn and future work is outlined.

Analysis of Rolling Friction for a Ball in a V-Grooved Track

As mentioned in the previous section, rolling friction may be caused by internal losses and/or external tractions. Consequently, it may be pertinent to gain some idea of the mechanisms, parameters, and orders of magnitudes of any possible mechanism prior to embarkation on a test program. In this section, we first examine the kinematics of motion of the ball in order to evaluate the creepages associated with that motion. Secondly, we assess the importance of each rolling friction mechanism.

Kinematics of a Ball Rolling in V-Grooved Tracks

As mentioned in the introduction, the tractive rolling subject of this investigation arises mainly as a consequence of creepages imposed on the otherwise pure rolling motion. For a formal definition of the principal creepages, the reader is referred to [6]. In the case of balls rolling between two V-grooved tracks, the most dominant type of creepage is spin. Quantification of relative spin, that is, the amount of spin per unit rolling distance is therefore essential for evaluating tractive friction. To this end, this section presents a kinematic analysis of the motion of the balls between the V-grooves.

Consider a ball rolling between two V-grooved tracks as depicted in Figure 1. When the lower track moves, with a velocity V normal to the plane of the page, the ball will roll between the grooves with a main rolling velocity ω. This latter may be resolved into a pure rolling motion ωR and a pure spin motion ωS. In the case of pure rolling, the creepage is assumed zero, thus no microslip takes place. In other words, for the case of a ball in the V-grooved tracks, the spin component of the rolling motion is the only creepage responsible for the frictional losses. As a consequence, the friction resistance will increase with decreasing groove angle, as the spin per unit rolling increases with a decreasing groove angle. The other limiting case is the rolling of a sphere in between two horizontal flat plates, where consequently no creepage occurs. In this case, the only remaining sources of rolling friction are elastic hysteresis losses and Heathcote slip (see further below).

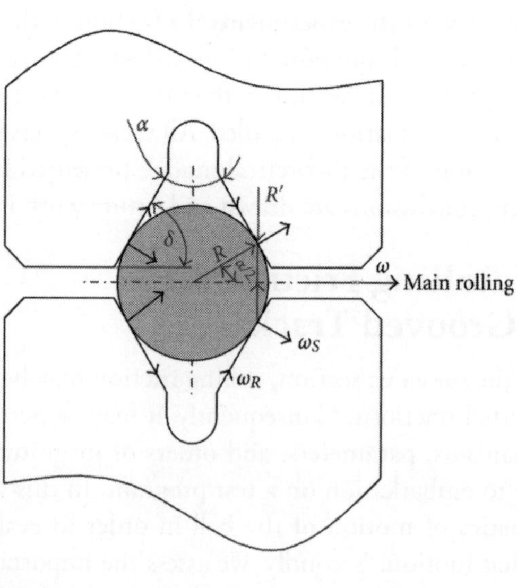

Figure 1: The roll and spin motion.

A linear motion of the lower V-groove corresponds to a main angular velocity ω:

$$\omega = \frac{V}{R'}, \qquad (1)$$

with R′ being the distance depicted in Figure 1. This main motion can be resolved into a pure rolling motion ω_R and a pure spin motion ω_S:

$$\omega_R = \omega\cos(\delta) = \omega\sin\left(\frac{\alpha}{2}\right),$$
$$\omega_S = \omega\sin(\delta) = \omega\cos\left(\frac{\alpha}{2}\right), \qquad (2)$$

$$R' = R\sin\left(\frac{\alpha}{2}\right),$$
$$\omega_S = \frac{V\cos(\alpha/2)}{R\sin(\alpha/2)} \qquad (3)$$
$$\omega_R = \frac{V}{R}.$$

Rolling Friction Forces

Beginning with the traction forces, let us note that the forces associated with the kinematics outlined above may be obtained in an analogous way. Thus, let the motion of the sphere result in a spinning moment Mz, at each contact patch, which gives rise to the frictional force that has to be overcome by the linear actuator. For one ball, four spin components Mz are formed, as depicted in Figure 2, which result in a global spin moment Mz, tot, around the rolling axis, that is given by

$$M_{z,tot}=4M_z\cos\left(\frac{\alpha}{2}\right). \qquad (4)$$

The rolling friction force, per ball, associated with this moment is given by

$$F_{\text{fric,tot}} = \frac{4M_z\cos(\alpha/2)}{2R'} = \frac{2M_z\cos(\alpha/2)}{R\sin(\alpha/2)} \qquad (5)$$

Thus, the coefficient of rolling friction owing to tractive rolling is given by

$$\lambda_{\text{tractive}} = \frac{F_{\text{fric}}}{W} = \frac{M_{z,\text{tot}}}{WR\sin(\alpha/2)}, \quad (6)$$

with W the externally applied normal load.

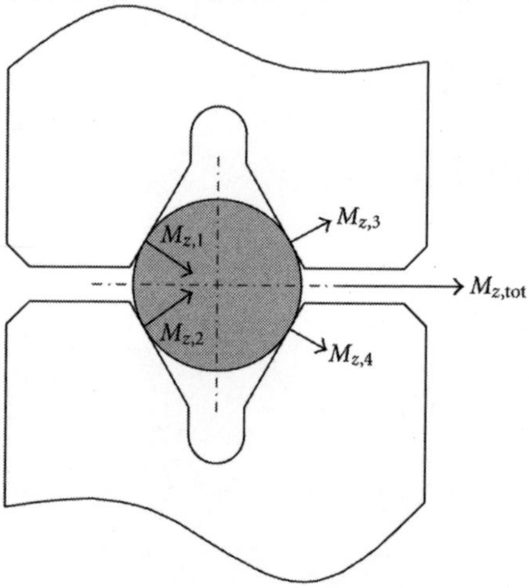

Figure 2: The local spin moments result in a global spin moment Mz, tot.

We turn our attention presently to the other two phenomena responsible for the measured friction in experiments, namely, bulk elastic hysteresis and Heathcote slip.

Firstly, rolling resistance as a consequence of the elastic hysteresis losses due to Hertzian compression is given by [13]

$$\lambda_{\text{elastic}} = \frac{3\epsilon a}{16R}, \quad (7)$$

where ϵ is the coefficient of elastic hysteresis loss due to Hertzian compression and a is the radius of the contact patch which is given by

$$a = \left(\frac{3}{4}\frac{WR}{E^*}\right)^{1/3}. \quad (8)$$

Secondly, Heathcote [14] deduced a formulation for the frictional moment M for the case of pure rolling by neglecting the elastic compliance of the contacting bodies [15],

$$\frac{M}{W} = 0.08\mu\frac{a^2}{R}, , \qquad (9)$$

with μ being the local coefficient of sliding friction. From this equation, one can derive an expression for the coefficient of rolling resistance due to Heathcote slip,

$$\lambda_{HC} = \frac{M}{RW} = 0.08\frac{\mu a^2}{R^2}. \ (10)$$

Order of Magnitude Analysis

The three phenomena together (6), (7), and (10) account for the total coefficient of rolling friction measured in the subsequently presented experiments. It is thus interesting to gauge the relative importance of these different rolling friction contributions in the context of our experiments.

Thus, let us consider a steel ball of 6 mm diameter, a maximal load W of 20 N and a coefficient of sliding friction μ of 0. 5. The contact radius a will be on the order of 10–4 m. Experiments reported in this paper show a rolling friction value on the order of 10–2. Considering a realistic coefficient of elastic hysteresis loss ϵ of 0. 01, the coefficient of rolling friction due to elastic hysteresis losses $\lambda_{elastic}$ is on the order of 10–5. The coefficient of rolling friction due to Heathcote slip λ_{HC} is on the order of 10–8.

This estimation indicates that the contribution of traction (in this case, microslip associated with the spin creepage) is by far the most dominant, compared to the two other energy dissipating phenomena.

Overview of the Experiments

The Test Setup

The experimental investigation is carried out on a previously developed tribometer (we refer to [3] for more details) with some minor adaptations. Figure 3 depicts a schematic view of the setup. The tribometer comprises four parts. (1) the actuator part is fixed to the base plate. It comprises a Lorenz actuator, which drives the lower friction member through a flexured rod. (2) The frictional part contains (i) the lower friction member, which in our case is the lower V-track mounted to a block that floats on air bearings and is driven by the aforementioned actuator, and (ii) the upper friction member which is the upper V-track, being connected to the dynamometer via an elastic hinge; (3) a normal loading part, which comprises a cantilever to apply the load through an air bearing; and (4) the sensor part which comprises (i) a Bently eddy current sensor which measures the relative motion between the two friction members, (ii) the friction force dynamometer which contains a force cell attached to a fixed frame that is fixed to the base plate, and (iii) a force cell for measuring the normal load. These different parts are decoupled as far as possible. The Lorenz actuator imposes a straight motion, with a periodic velocity, to the lower friction member through the flexured rod. The Lorenz actuator is current driven and the position signal of the displacement sensor is used for feedback control to obtain, as best as possible, the desired motion of the moving part.

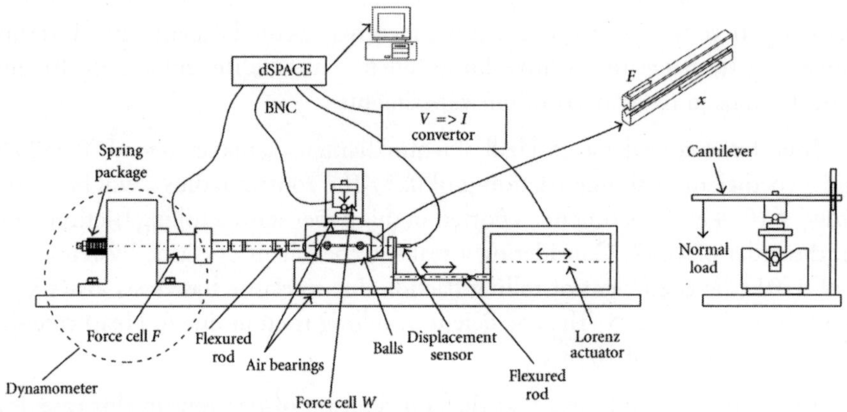

Figure 3: A schematic overview of the tribometer.

The purpose of the employed elastic joints is to set off alignment errors of the friction block in the normal, lateral, and rotational directions.

The loading part allows easy control of the normal load on the frictional part. The load is applied by using a cantilever (see Figure 3) and is transmitted to the frictional part through an air bearing, in order to avoid affecting the inertia of the frictional part when changing the normal load (mass). Another advantage of the air bearing is that the load is applied without direct contact which may introduce adventitious forces. The air bearing is equipped with a ball joint to allow the bearing to adapt its orientation with respect to the friction block. In this way, a uniform load is applied normal to the friction contact. Detailed experimental characterization of this setup can be found in [16].

The Normal Load Limitations

In order to guarantee the applicability of Hertz's theory and the elastic behavior of the contacting bodies, the maximum allowable load is calculated. From the work of Tabor [17], plasticity first occurs when the Hertzian contact pressure reaches the limit of approximately 0. 6H, where H is the hardness. Thus for a circular contact,

$$p_0 = \frac{1}{\pi}\left(\frac{6WE^{*2}}{R^2}\right)^{1/3} < 0.6H. \quad (11)$$

In Figure 4, the evolution of the contact pressure is depicted as function of the normal load on the V-grooved track for different opening angles of the V grooves and the case of two steel balls of diameter 6. 35 mm. For the used V groove of 60°, the maximum applicable load approximates to 200 N. For higher loads, the brinelling effect occurs which means that plastic damage of the contacting surfaces takes place.

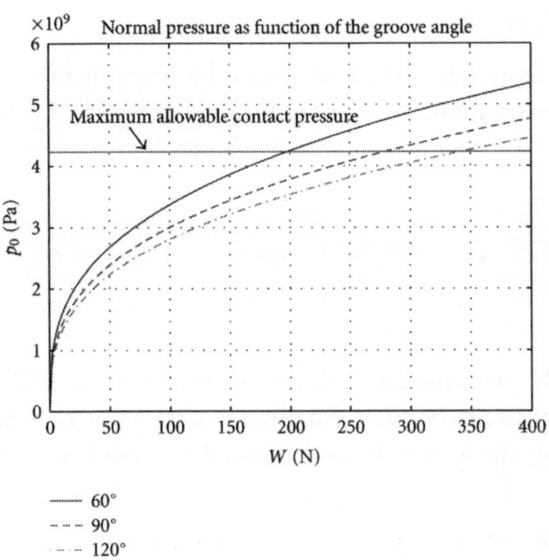

Figure 4: Determination of the maximum allowable normal load (6. 35 mm steel balls).

Measurement Procedure

For the purpose of this study, the displacement of the moving block and the friction force are measured by the sensors described above and captured using a dSPACE system. To improve the quality of the measured data, a lowpass filter with a bandwidth of 30 Hz is used to exclude most of the measurement noise. The filter is applied to all channels in the same way in order to avoid phase shift effects. Afterwards, the data samples, obtained from repeated test runs for any single test condition, are time averaged to remove random noise and to improve their statistical relevance.

Identification of Prerolling Hysteresis

The aim of the measurements is to evaluate the prerolling behavior of the friction force for different normal loads. The prerolling friction force is function of the displacement which can be described by a rate-independent hysteresis function. The hysteresis curve is fully determined by its virgin curve (see [18] for detail). This theoretical curve starts at the origin (with the very first commencement of motion) has positive first and second derivatives for all displacements, and saturates for large displacements. Therefore, it is interesting to use the virgin curve as reference for comparing the hysteresis curves for the different experiments, since this curve incorporates all information of the hysteresis behavior. Once a hysteresis loop is identified, one can determine the virgin curve by scaling the identified curve, by factor 2 in both x and y directions, and shifting the turning points to the origin [18].

A measured hysteresis curve is identified by fitting it to a parametric model using nonlinear regression, in order to capture the curve in a mathematical form. In this study, a two parameter model is chosen for simplicity which is of the form

$$F(x) = h_0\left(1 - e^{-a(x-x_0)}\right). \quad (12)$$

The parameter h_0 gives the saturation value of the curve, while the parameter a is a measure of the curvature. Finally, the parameter x_0 is necessary for the translation along the x-axis. For other suitable model equations, we refer to [2].

Figure 5 shows an example of the model equation fitted to measured hysteresis data. Both the external loop and the derived virgin curve are depicted.

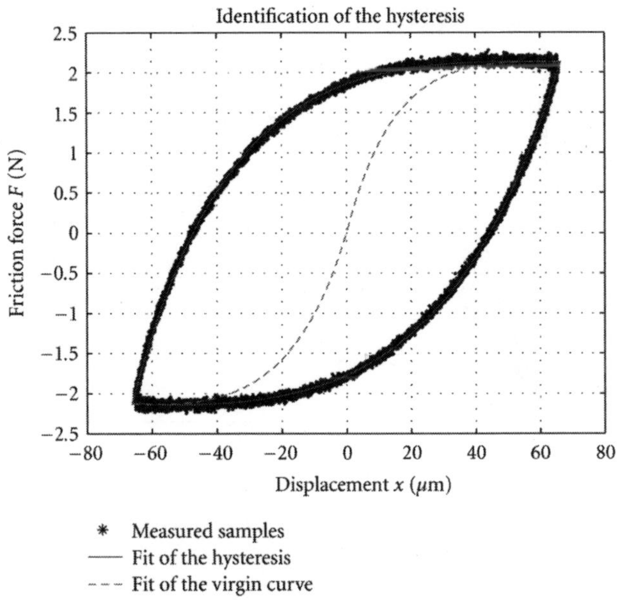

Figure 5: Identification of the hysteresis loop, including the determination of the virgin curves. (Steel balls ⌀6. 35 mm, groove angle: 60°, W=88. 9 N).

Influence of the Normal Load

Examples of the fitted virgin curves for the different normal loads are visualized in Figure 6. The determining parameters for these curves are h_0 and a which are depicted in Figures 7(a) and 7(b). One can observe an increasing trend for h_0 as function of the external normal load W, while the curvature a as function of the external normal load W shows a decreasing trend. Both trends can be fitted by a linear equation, as shown in the figures. The range of performed experiments result in a spread of h_0 of approximately factor 3 (between minimum and maximum values), while the results for parameter a give a spread of approximately a factor of 1. 1. For the sake of gaining an idea about the shape deviations, let us take a to be fixed to a value equal to its mean value of 0. 0325. In that case, the error on a approximates to ±5%. In this study however, the parameter a is considered to have a linear decreasing trend as function of the normal load W which reduces the error further to approximately 3%. In other words, the effect of the normal load on the shape of the hysteresis curve is not appreciable.

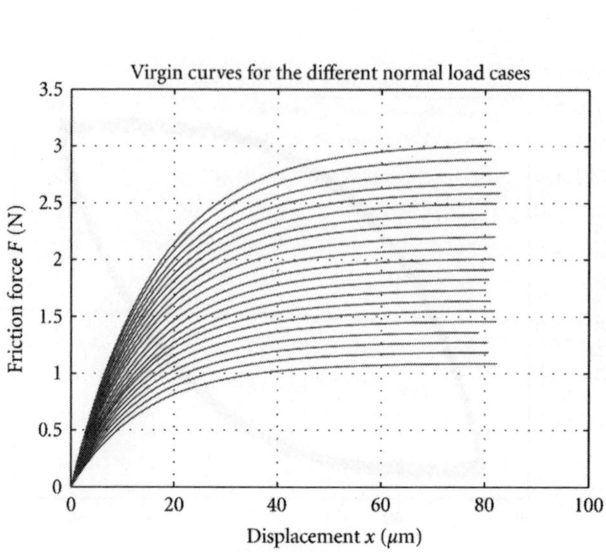

Figure 6: The identified virgin curves for the different normal load cases (46. 5 N≤W≤123. 2 N).

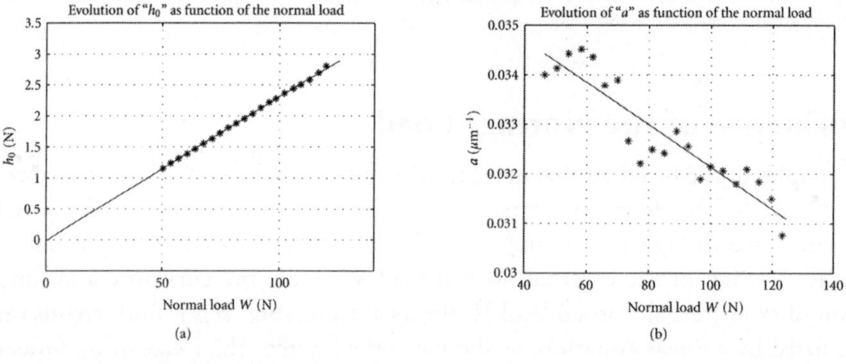

Figure 7: The evolution of h_0 (a) and a (b) with the normal load.

In Figure 8, the virgin curves are scaled by the applied normal load to obtain the evolution of λ=F/W, the coefficient of rolling friction. One can observe a slightly increasing trend for λ in function of the external normal load.

In order to ascertain the influence of sliding friction characteristics (of the ball/guideway surfaces) on the rolling friction values, experiments with other ball types are carried out on the same setup. Balls of diameter of 6 mm, coated with a thin layer of diamond-like carbon, are compared to the former uncoated steel balls. The results are depicted in Figure 9. While the same general trends are visualized for these DLC coated balls, one can notice the appreciably lower rolling friction values (by factor 0. 3) for the coated balls. This confirms our previous conclusion (see Section 2. 2) that the greatest part of the rolling resistance, for this bearing configuration, is owing to microslip tractions and not to elastic hysteresis in the ball/groove material. Furthermore, for the DLC-coated balls, the spread of the parameter h0 approximates to a factor of 2. 8. As for the parameter a, the error committed by taking it as a constant (equal to 0. 125), is ±12%, while considering the linear decreasing trend reduces the error to approximately 1%.

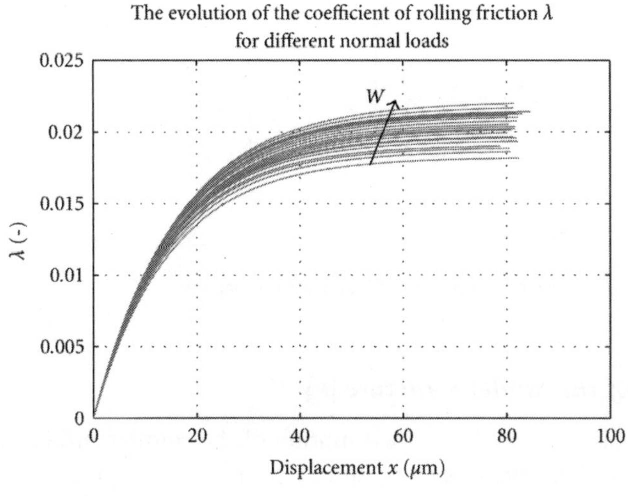

Figure 8: The evolution of λ with the normal load.

Comparison Between Model and Experiment

The second aim of this article is the validation of a theoretical model presented in [6]. The model is able to predict creepage induced traction during and after prerolling.

For the sake of completeness, a short overview of the model is given in the next section. Afterwards, the model is validated by means of the previously presented experimental results for the case of rolling with spin.

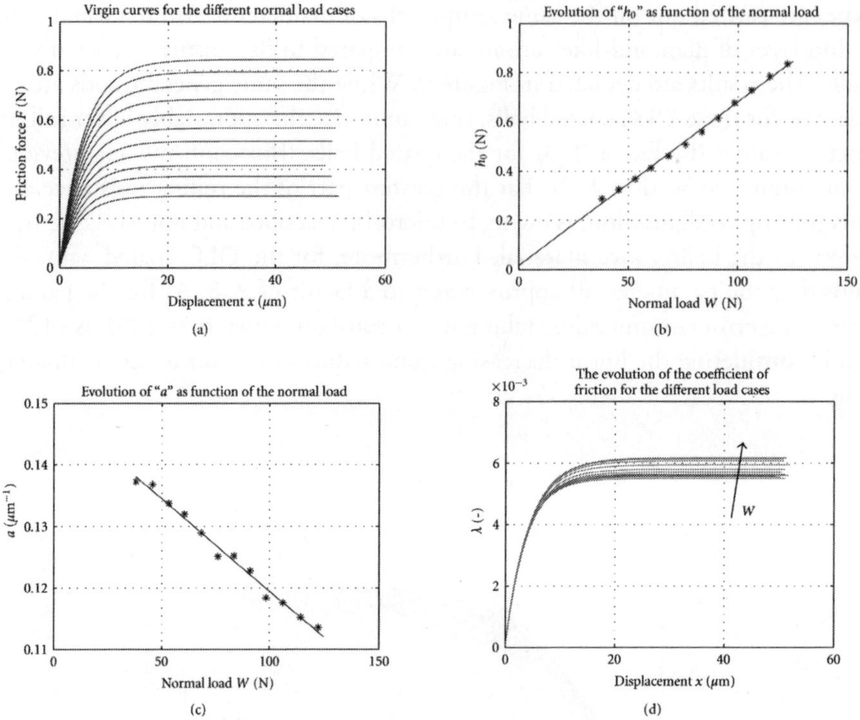

Figure 9: Results for the diamond coated balls (⌀6 mm, 60° opening angle).

Overview of the Model Structure [6]

For rolling objects, the relative rigid motion of the counter surfaces termed the creepage, C, which consists of longitudinal and lateral creepages, and spin. The (elastic) surface displacement field, u, is related to the slip field, S, by the equation

$$\mathbf{S} = \mathbf{C} - \frac{\partial \mathbf{u}}{\partial x} + \frac{\partial \mathbf{u}}{\partial q}, \quad (13)$$

Analysis for the case of no slip results in closed form expression for the displacement of the surface particles:

$$u = f(C, x, y, q), \quad (14)$$

where x, y are the coordinates of the points in the contact patch, and q is the traversed rolling distance.

To take slip into consideration, the newly obtained values u=(u, v) are used to calculate the pressure distribution in the contact patch:

$$(p_x, p_y)=\mathrm{L}(u, v). \quad (15)$$

If $\|(p_x, p_y)\| \leq \mu p_z$, no slip occurs ($S_x=S_y=0$) and the values of (u, v) are retained. In the other case, $\|(p_x, p_y)\| > \mu p_z$, the no slip deformation is replaced by

$$(u, v)_{\text{new}} = \frac{\mu p_z}{L}\left(\frac{(u, v)}{\|(u, v)\|}\right)_{\text{old}}. \quad (16)$$

Finally, the traction forces and spin moment are calculated as follows:

$$\begin{aligned}(F_x, F_y) &= \int_{-a}^{a}\int_{-a(y)}^{a(y)}(p_x, p_y)\,dx\,dy,\\ M_z &= \int_{-a}^{a}\int_{-a(y)}^{a(y)}p(x, y)r(x, y)\,dx\,dy.\end{aligned} \quad (17)$$

In order to calculate the tractions using this model, one needs to know (i) the geometry and elastic properties of the rolling objects, (ii) the normal load, (iii) the creepages, and (iv) the local sliding friction coefficient in the contact. From the kinematics of a V-grooved setup, it appears that only spin creepage is present. Furthermore, only the spin moments are of interest, since the force in the rolling direction F_x equals zero for every contact patch, and the force, Fy, perpendicular to the rolling direction will have a resultant that is equal to zero when taking all four contact patches into consideration.

Experimental Validation of the Model

To validate the model, presented in [6], the experimental results are compared to the model predictions. The model is implemented for the kinematics of the V-grooved track, using reasonable values for the elastic parameters. The local coefficient of sliding friction, necessary for the model, is determined by an additional experiment. In this experiment, the coefficient of sliding friction is determined for the system of a ball on the track as sliding counter surface. This is carried out by gluing two balls on a plate, and using this as counter surface for one of the V-grooves. Periodic sliding motion is imposed for different normal loads, while the friction force is measured. The coefficient of sliding friction is determined by dividing the measured friction force by the applied normal load. Although this rudimentary experiment ignores some basic rules as, for example, the velocity dependence of the coefficient of friction; these estimates are accurate enough for a first evaluation of the model and are a good starting point for further analysis. Note that, Lampaert et al. [3] show that the velocity dependence of the coefficient of friction for metallic contacts is small. The sliding friction results of this experiment are depicted in Figure 10 for different normal loads. The results of this short experiment are used as input for the model.

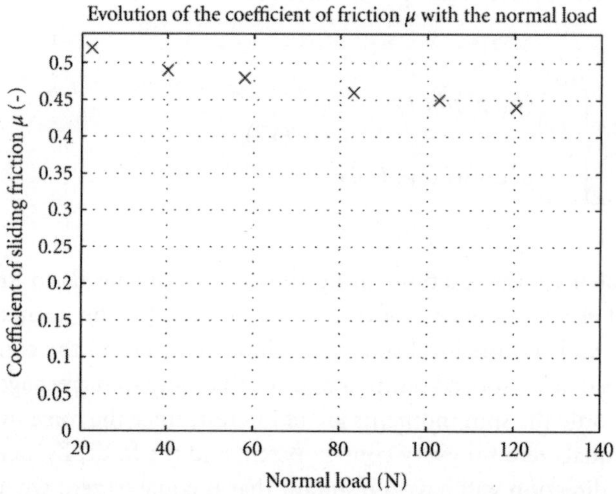

Figure 10: The evolution of the coefficient of sliding friction as function of the normal load.

The experimentally obtained virgin curves of Section 4 are now compared with the model results in Figure 11. We note a good resemblance between model and experimental results. The initial slope as well as the saturation value behave in ways similar to theoretical predictions.

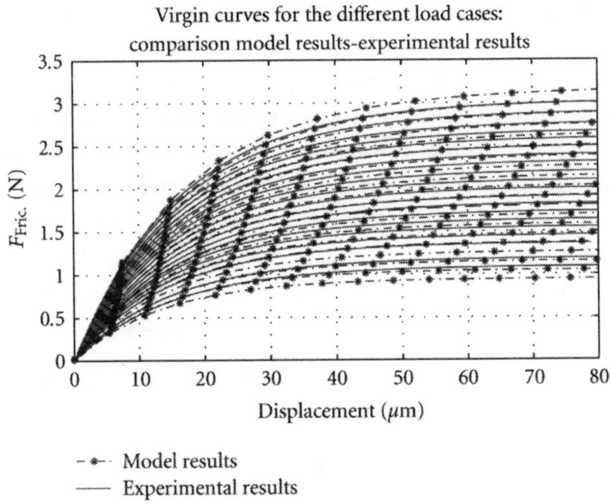

Figure 11: A comparison between experimental results and model results of [6] (6. 35 mm steel balls, 60° groove angle, 46. 5 N≤W≤123. 2 N).

In Figure 12, the error, according to (18) below, between model and experiment is presented for the different virgin curves. For clarity, only a few characteristic results are presented. The error is defined as follows:

$$\text{error} = \frac{|(F_{\text{measured}} - F_{\text{model}})|}{F_{\text{mean}}}, \quad (18)$$

where $F_{mean}=(F_{measured}+F_{model})/2$, calculated as function of the displacement.

In Figure 13, for every virgin curve, the error is averaged and divided by the maximum friction force. One can distinguish two regions. Firstly, experiments with high normal loads show an error lying within 5%. Secondly, experiments with small normal loads show an error lying in the range of 6–16%. An obvious explanation to this is the smaller signals and thus the lower signal-to-noise ratio of those experiments.

Discussion and Conclusions

In this paper, an experimental investigation of prerolling is presented. More specifically, the influence of the normal load on the hysteresis behavior is studied on an existing setup, in which the frictional element comprises two balls rolling between a pair of V-grooved tracks. The frictional hysteresis is fitted by an exponential model form, in which two parameters characterize the shape, denoted by the curvature, and the saturation value of the friction force, respectively. The evolution of these parameters as function of the normal load is presented. We observe a slight increase in the coefficient of rolling friction λ with an increasing normal load and a very small change in the curvature.

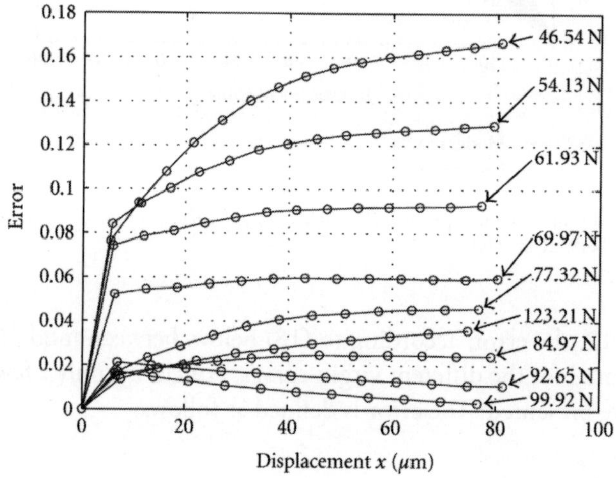

Figure 12: The error between model and experimental results.

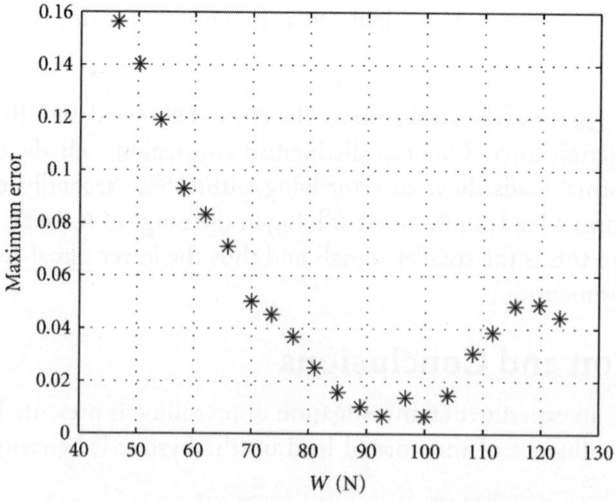

Figure 13: The maximum error between model and experimental results as function of the normal load.

Another part of the investigation carries out a validation of an existing theoretical model for tractive rolling by use of the obtained experimental data. Kinematic analysis of a rolling configuration shows that the only creepage present is spin. Using the experimentally determined sliding friction coefficient and spin values as input to the model, a comparison of model prediction and experiment is carried out. Agreement between model and experiment falls within an error margin of less than 16% for low normal loads and 5% for high normal loads.

Future Work

The global aim of this experimental research is the determination of the relation between the hysteretic friction force and the different system parameters. In [2], Al-Bender and Symens investigate other system parameters as groove angle, ball diameter, and so forth. A global synthesis of these results, together with the results of this study, can improve the understanding of the frictional behavior of rolling elements. A more correct estimation of the hysteresis shape can find its applications in feed forward position control to avoid more complicated model structures which are computationally intensive and less appropriate for online application.

Acknowledgments

This research is supported by the fund for Scientific Research-Flanders (F. W. O.) under Grant FWO4283. Professor P. Vanherck is greatly acknowledged for fruitful discussions and comments. Tegoeh Tjahjowidodo and Thierry Janssens are acknowledged for their support during the experiments. The scientific responsibility is assumed by its authors.

References

1. B. Armstrong-Hélouvry, Control of Machines with Friction, Kluwer Academic Publishers, Norwell, Mass, USA, 1991.
2. F. Al-Bender and W. Symens, "Characterization of frictional hysteresis in ball-bearing guideways, " Wear, vol. 258, no. 11-12, pp. 1630–1642, 2005.
3. V. Lampaert, F. Al-Bender, and J. Swevers, "Experimental characterization of dry friction at low velocities on a developed tribometer setup for macroscopic measurements, " Tribology Letters, vol. 16, no. 1-2, pp. 95–106, 2004.
4. F. Al-Bender and W. Symens, "Dynamic characterization of hysteresis elements in mechanical systems—I: theoretical analysis, " Chaos, vol. 15, no. 1, Article ID 013105, 11 pages, 2005.

5. F. Al-Bender and W. Symens, "Dynamic characterization of hysteresis elements in mechanical systems—II: experimental validation, " Chaos, vol. 15, no. 1, Article ID 013106, 9 pages, 2005.
6. F. Al-Bender and K. De Moerlooze, "A model of the transient behavior of tractive rolling contacts, " Advances in Tribology, vol. 2008, Article ID 214894, 17 pages, 2008.
7. Q. Y. Liu, X. S. Jin, and Z. R. Zhou, "An investigation of friction characteristic of steels under rolling-sliding condition, " Wear, vol. 259, no. 1–6, pp. 439–444, 2005.
8. K. G. Budinski, "An inclined plane test for the breakaway coefficient of rolling friction of rolling element bearings, " Wear, vol. 259, no. 7–12, pp. 1443–1447, 2005.
9. H. Endo, E. Marui, N. Hasegawa, and H. Katagiri, "Transient rolling behaviour of a slider supported by two rollers, " Tribology International, vol. 31, no. 7, pp. 377–384, 1998.
10. J. Otsuka and T. Masuda, "The influence of nonlinear spring behavior of rolling elements on ultraprecision positioning control systems, " Nanotechnology, vol. 9, no. 2, pp. 85–92, 1998.
11. J. Wang, T. Hashimoto, H. Nishikawa, and M. Kaneta, "Pure rolling elastohydrodynamic lubrication of short stroke reciprocating motion, " Tribology International, vol. 38, no. 11-12, pp. 1013–1021, 2005.
12. L. Xiao, S. Björklund, and B. G. Rosén, "The influence of surface roughness and the contact pressure distribution on friction in rolling/sliding contacts, " Tribology International, vol. 40, no. 4, pp. 694–698, 2007.
13. R. D. Arnell, P. B. Davies, J. Halling, and T. L. Whomes, Tribology Principles and Design Applications, MacMillan, London, UK, 1991.
14. H. L. Heathcote, "The ball bearing: in the making, under test and on service, " Proceedings of the Institute of Automotive Engineers, vol. 15, pp. 569–702, 1921.
15. K. L. Johnson, Contact Mechanics, Cambridge University Press, Cambridge, UK, 1985.
16. V. Lampaert, Modelling and control of dry sliding friction in mechanical systems, Ph. D. thesis, Division of Production Engineering, Machine design and Automation, Katholieke Universiteit Leuven, Leuven, Belgium, 2003.

17. D. Tabor, The Hardness of Metals, Oxford University Press, Oxford, UK, 1951.

18. F. Al-Bender, W. Symens, J. Swevers, and H. Van Brussel, "Theoretical analysis of the dynamic behavior of hysteresis elements in mechanical systems, " International Journal of Non-Linear Mechanics, vol. 39, no. 10, pp. 1721–1735, 2004.

Monitoring of Diamond Mill Wear in Time Domain During Stone Cutting Using Cutting Force Measurements

W. Polini and S. Turchetta

ABSTRACT

The main focus of this study is to verify the possibility to monitor diamond mill wear by measuring the cutting force during stone machining. An ornamental stone was machined on an NC machining centre, retrofitted with a 3-axis dynamometer and data acquisition systems, to investigate the effects of variations in machining parameters. The sensor data include cutting force measurements, further divided into measurable components, such as x, y, and z. Those components were analysed to determine the sensory features that best correlate with diamond mill wear. Tool wear and machining parameters appeared significantly influencing cutting force signals. This suggests that even under the varying cutting conditions involving different values of process parameters, the identified cutting force feature can be used for the reliable and accurate control of diamond mill wear during stone cutting operations.

Introduction

The intense international market competition in stone field has driven the attention of manufacturers on the automation of manufacturing systems as means for increasing the productivity and improving the quality of products.

Monitoring of the stone cutting processes plays a very important role to avoid down time of the machine, or to prevent unwanted conditions such as excessive tool wear or breakage. A recent survey of literature in the field of tool condition monitoring stresses the fact that the choice of sensors and the signal interpretation are critical to establish a reliable tool condition monitoring system [1]. This is due to the fact that manufacturing processes are difficult to model mathematically. Sensors supply the information that we need to describe the process.

A vast amount of literature exists in the field of tool wear monitoring in metal cutting environment; it suggests that a variety of process parameters can be used to predict the cutting tool state [1]. Techniques for online wear monitoring can be grouped into two main categories: direct sensing and indirect sensing techniques. When direct methods of wear measurements have been attempted, the majority of methods have been indirect. Direct methods are less beneficial because the cutting area is largely inaccessible, and therefore, online monitoring cannot be carried out while the tool is engaged in in-process cutting. These methods include touch trigger probes, optical, radioactive, proximity sensors, and electrical resistance measurement techniques. Indirect methods take measurements while the tool is actively engaged, since it involves recording a variable that can be correlated to tool wear. Commonly used methods include acoustic emission [2], tool temperature, cutting force [3, 4], vibration signature [5–10], and miscellaneous methods, such as ultrasonic and optical measurements, workpiece surface finish quality [5], workpiece dimensions, stress/strain analysis, and spindle motor current. The cutting conditions (speed, feed-rate, and depth of cut) are known to affect the sensor signals and a range of methods have been suggested for separating the effects of these conditions from those of wear on the measured parameter.

Methods for correlating the measured process parameters to tool wear, breakage, or chipping fall primarily into three categories [11]. The first class consists of methods that can be viewed as heuristic-based rules with a priori knowledge only of the process parameters, such as mathematical modelling and adaptive observers. The second category of methods is that requiring formal knowledge of the process and can be grouped together as analytical-based models such as time series analysis and fast Fourier transformation (FFT). The last category is one of the example-based models with inductive learning capabilities such as pattern recognition, decision surfaces, mapping techniques, clustering, and artificial intelligence models (neural networks, genetic algorithms, and knowledge-based systems) [12].

The present work shows a method for tool condition monitoring of a diamond mill used in stone cutting. It focuses on synthetic diamond tool which is a fairly expensive tool used commonly in the cutting of ornamental stone. A work exists on tool condition monitoring of this kind of tool in [13]. It implements a monitoring system utilising simultaneous vibration and strain measurements on the tool tip for the wear of synthetic diamond tool which are specifically used for the manufacturing of aluminium pistons.

This work presents an approach to investigate the correlation of tool wear and cutting force. The idea is to establish tool wear progression by monitoring cutting force during stone machining. It investigates the feasibility of using the cutting force signatures to monitor tool wear. The characteristics of the cutting force signatures were examined in time domain. The identification of common sensory features that best correlate tool wear condition with tool wear was adequately studied. To address the problem, stone cutting on Coreno Perlato Royal by an NC machining centre retrofitted with force sensors and data acquisition systems was conducted to find the relationship among force signal and tool wear. Different settings of cutting parameters were intended to simulate diverse processes. Each measurement was further divided into components: x, y, and z cutting forces (Fx, Fy, and Fz). Those components were processed for time domain features. Using the statistical analysis techniques, some components of sensory data were found to be the most influential factors to tool wear. Therefore, the identified sensory features can be used for the control of diamond mill wear in stone cutting more reliably and accurately under varying cutting conditions.

The objectives of this paper are (1) to define what is the wear of a diamond mill, (2) to present the method to monitor the tool condition during stone cutting, and finally, (3) to validate the proposed approach by means of experiments.

Wear of a Diamond Mill During Stone Cutting

Stone cutting by means of diamond tools consists in wearing away its mineral constituents by passing hard grits over the machined surface. A diamond mill is composed of a support, abrasive grits, and a bond. The support is that part of the tool which the abrasive grits are fixed to; it gives the right shape to the tool; it transmits the kinetic energy from the machine axes to the abrasive grits; it absorbs the stresses generated during machining. The abrasive grits constitute the cut edges; they are characterised by a granulometry, a shape, and a concentration. The bond is a metallic alloy blocking the abrasive on the tool support in order to make the machining possible. It assures both cutting ability and long tool life. A diamond mill for stone cutting is electroplated, that is, it is constituted by a layer of diamond grits fixed to the mill support by the bond.

Many works in literature assume as tool failure mode the flank wear, the fracture, the crater wear, and the plastic deformation [13]. Flank and crater wear are generally accepted as the normal tool failure modes. In the case of synthetic diamond tools used for this study, the flank wear did not play a very important role.

The wear of an electroplated diamond mill is the result of the progressive consumption of diamond grits [14]. It is characterised by a sequence of steps, as shown in Figure 1 [15]. Firstly, tool surface is characterised by whole crystals that are diamond grits having a large height of protrusion and a minimum surface damage. Once the workpiece is contacted, the grit is rounded off so much that a plateau is generated on the grit top. It generates mechanical friction and thermic effects that polish the diamond grits. A high percentage of polished grits makes the surface of the tool so vitreous, to cause a less efficient cutting ability. The intermittent contact with the workpiece, caused by the tool rotation, leads to a cyclic load on the diamond grit which deteriorates the grit ability to contrast the cutting forces, and finally, the grit breakage. This phenomenon is enlarged by the heterogeneous nature of the workpiece and by the machining vibrations. The diamond grits result is characterized by microfractures and microcracks that increase the number of cut edges on grit surface, even if they reduce the penetration depth. Wear progression leads to a completely fragmentary particle till the tool looses its cutting performances.

Figure 1: Wear of an electroplated diamond mill: (a) protruded grit, (b) flattened grit (c) microfractured grit, and (d) macrofractured grit.

The progression of diamond tool wear is carried out to consume the diamond. The microgeometric effect is both the progressive rounding and breakage of the diamond grits that are placed in a unique layer on mill surface. In previous works, we have presented a test protocol [16] to be implemented in order to measure the microgeometric parameters [17] that characterise the wear of a diamond mill. This protocol implies the need to observe the tool, once dismounted by the mandrel, after the cut. The present paper overcomes the need to dismount the tool by the mandrel by implementing an online measurement of diamond mill wear.

Monitoring Strategy

Experimental Setup

In stone machining, the formulation of accurate mathematical models is very difficult, hence in most cases, empirical equations are obtained by performing statistical analysis or through the training of neural network models to fit the experimental data. To collect the data, a designed experiment is performed using predetermined values of machining parameters. Even though empirical equations do not consider the underlying principles and mathematical relationships, empirical modelling does have advantages of accuracy and ease of formulation. Once noticeable, drawback of any empirical modelling is the rigidity that allows only limited variations in the machining conditions. That is, the cutting conditions must be carefully controlled to remain similar to those used in the cutting experiments. Development of empirical models often leads to a large number of trials and a considerable amount of time and effort, however, whenever variations occur, empirical models need to be refitted to maintain the accuracy. To avoid costly reproduction of cutting experiments under the diverse production environment, while satisfying the quality characteristics of the machined products, reliable and accurate means of tool condition monitoring and process control become necessary.

In this study, all machining experiments were carried out on an NC machining centre of Brembana Macchine. The experimental setup is shown in Figure 2. A diamond mill, commonly used to contour ornamental stone, was used, with a diameter and a height of 20 mm and 37. 5 mm, respectively. The considered diamond mill has a diamond concentration of 4. 77 diamonds/cm^2 and a diamond grit size of 40/50 mesh. Its bond is constituted by 90% of nichel.

The workpiece material was Coreno Perlato Royal marble. It mainly consists of CaCO3 with inclusions of seaweed and fossils that produce light and dark spots appreciated from an aesthetic point of view. Its mechanical properties are shown in Table 1 in comparison with those of well-known white carrara marble.

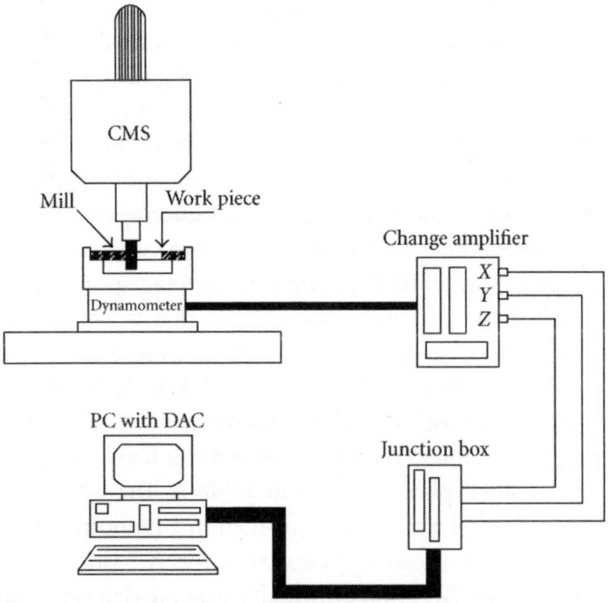

Figure 2: Experimental scheme.

Table 1: Mechanical properties of Coreno Perlato Royal and white carrara marble.

Material properties	Coreno Perlato Royal	White carrara
Density [kg/m^3]	2740	2705
Water absorption [%]	4.0	0.06
Compressive strength [MPa]	163	131
Compressive strength to ageing by frost [MPa]	162	126
Young modulus [MPa]	72000	75000
Flexural strength [MPa]	12,8	16.9
Abrasion resistance	0.95	0.52
Impact resistance [cm]	32	61
Knoop hardness [MPa]	2001	1463

Twenty diamond grits were identified on the surface of the new diamond mill and their protrusion height was measured by a Leica microscope. Then, the new tool was mounted on the mandrel of the CNC machining centre in order to execute a linear cut inside a slab of Coreno Perlato Royal of 20 mm thickness, as shown in Figure 3. The cutting force signal was acquired along the whole cut. Then, the wear of the mill was accelerated by carrying the tool to execute a series of cuts on a slab of a very abrasive stone, the Sardinian granite. Once reached a certain stage of wear, that was identified by measuring the protrusion height of the considered 20 diamond grits on mill surface, the mill was carried to execute a linear cut on the slab of Coreno Perlato Royal once again. This sequence was repeated for five times, since the diamond mill wear was monitored in 5 steps distributed along the whole mill life. The mill life corresponds to the time during which the mill shows cutting performances, that is, the grit passes from a whole crystal to a fragmentary particle. Three feed speed values u (200, 400, and 600 mm/min) and three cutting depth values p (0. 1, 0. 5, and 1 mm) were taken into account; they were chosen in order to reproduce the commonly used industrial range of process variables. The spindle speed was fixed to 3000 rotations per second. The experimental plan is shown in Table 2. The whole number of test was 135. The experimental cuts were performed in a random sequence, in order to reduce the effect of any possible systematic error. A Kistler three-component force dynamometer (type 9257BA) on which the stone slab was mounted sensed the dynamic cutting force components (F_x, F_y, and F_z), as shown in Figure 3. The signal acquired by the dynamometer was sampled at different frequencies and for different time intervals in order to set the acquisition parameters giving the whole information about force signal with the minimum time waste. The force signal along x, y, and z axes was periodic and 8000 acquisition points seem to be enough to keep the whole signal information. Therefore, the output of the dynamometer was fed into an A/D converter and sampled at 10 kHz by a PC. Each observation consisted of about 0. 8 129-second time signal. The raw cutting force data were transformed into time domain.

Table 2: Experimental plan. Figure 3: Stone linear cuts.

Factors	Number of levels	Levels
Cutting depth (mm)	3	0.1, 0.5, 1.0
Feed speed (mm/min)	3	200, 400, 600
Tool wear	5	1, 2, 3, 4, 5
Spindle speed (rpm)	1	3000
Replications	3	

Cutting Force Signal Analysis in The Time Domain

A monitoring strategy was adopted based on time-domain characteristics. Figures 4(a)–4(c), 5(a)–5(c), and 6(a)–6(c) show typical time domain signals monitored in the X (transverse to feed), Y (feed), and Z (vertical) directions, respectively, at different time intervals inside tool life. Stone cutting was performed at a feed speed of 600 mm/min and at a depth of cut of 1 mm.

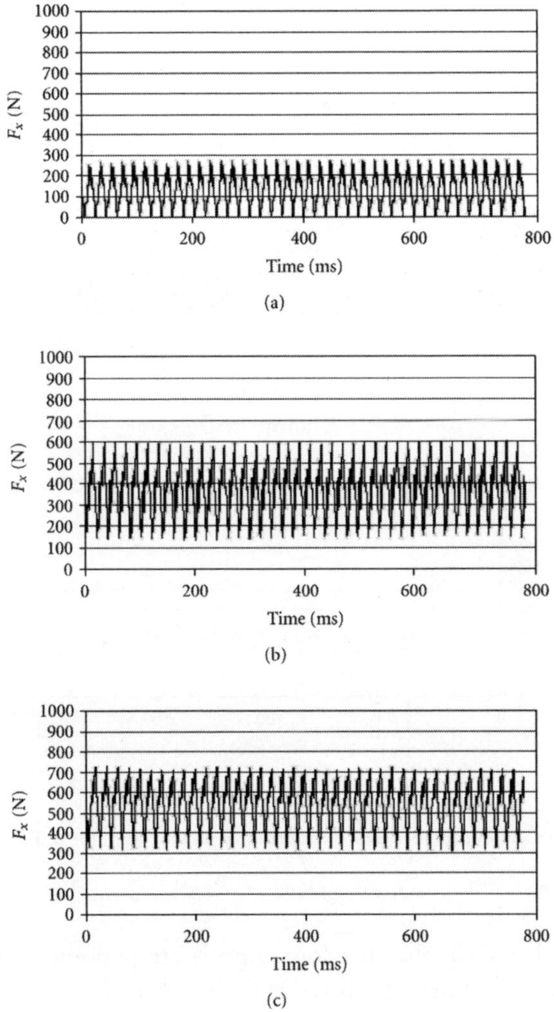

Figure 4: Time domain signal monitored in X direction: (a) at 1° wear stage, (b) at 3° wear stage, and (c) at 5° wear stage.

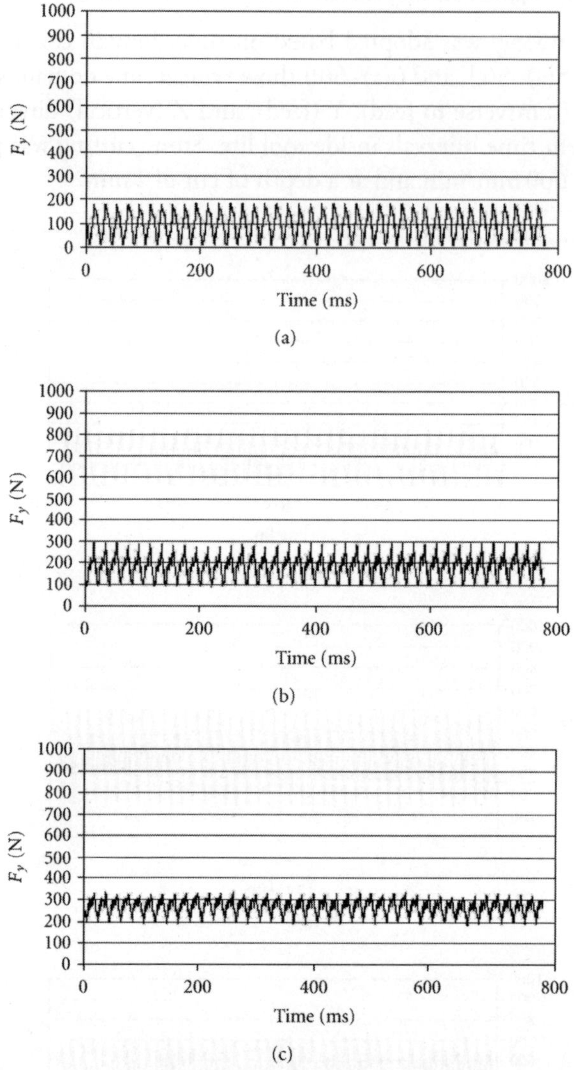

Figure 5: Time domain signal monitored in Y direction: (a) at 1° wear stage, (b) at 3° wear stage, and (c) at 5° wear stage.

Figures 4 and 5 show that the force signals are periodic and regular. The waviness of the cutting force signals is due to the eccentricity of the tool. In fact, many further tests have been carried out at 70–80 Hz. They have shown that the main frequency of the signal is related to the spindle speed of the tool. The average and maximum values of F_x and F_y signals increase with the progress of mill wear. The minimum value of F_x and F_y signals is near to 0 when mill is new, this is due to the fact that the cut is interrupted. In fact, the diamond grits on mill surface protrude by bond with different heights; this means that only the higher grits engage stone for cutting. When a grit exits stone, it is possible that the following grits do not engage stone, since it does not protrude enough by bond, causing the force on stone to reduce to zero. However, the average value of the F_z signal is equal to zero in each of the three considered cases, since the tool axis is perpendicular to the x-y plane, that is, a plane cut has been carried out. Therefore, F_z results have not been investigated further.

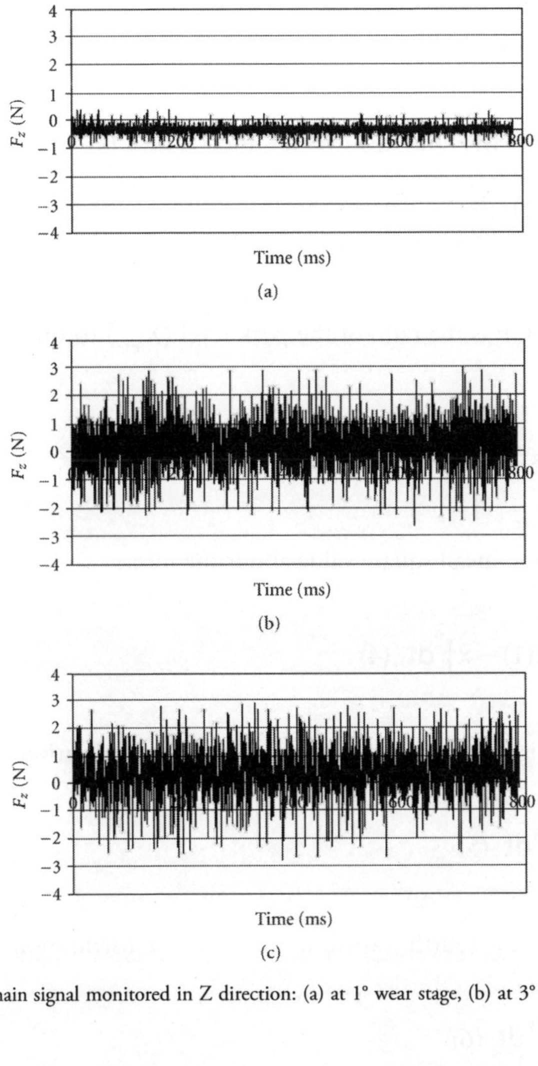

Figure 6: Time domain signal monitored in Z direction: (a) at 1° wear stage, (b) at 3° wear stage, and (c) at 5° wear stage.

After the signal was stored, a feature extraction program in Labview environment starts analysing the data. Previous studies have shown that some time-domain features can be indicative of tool wear for metal machining [1–4]. Therefore, the following time-domain features were extracted from each of the following parts: mean, root mean square, crest factor, variance, skewness, and kurtosis, in order to understand which is the time-domain feature that feels the main effects of diamond mill wear during stone cutting. These 6 features were extracted for the two force signals, F_x and F_y, respectively.

The mean value $\bar{x}$ of a function x(t) over an interval *T* is

$$\bar{x} = \frac{\int_0^T x(t)dt}{T}. \quad (1)$$

The X_{rms} value of a function x(t) over an interval of T is

$$X_{rms} = \sqrt{\frac{\int_0^T x(t)^2 dt}{T}}. \quad (2)$$

The crest factor CF is the ratio of the peak level (X_{max}) to the root mean square level (X_{rms})

$$CF = \frac{X_{max}}{X_{rms}}. \quad (3)$$

The variance is the mean square value about the mean

$$\sigma^2 = \frac{1}{T} \cdot \int_0^T [x(t) - \bar{x}]^2 dt. \quad (4)$$

The skewness S is the third statistical moment of a distribution

$$S = \frac{1}{\sigma^3 T} \int_0^T x^3 dt. \quad (5)$$

The kurtosis K is the fourth statistical moment of a distribution

$$K = \frac{1}{\sigma^4 T} \int_0^T x^4 dt. \quad (6)$$

The analysis of variance (ANOVA) was carried out on the 6 features previously defined in order to measure which feature feels the most significant influence of mill wear. The ratio between the variance of each feature data due to mill wear and the whole variance of each feature data was used to measure the influence of mill wear on each of the 6 considered features. Moreover, ANOVA analysis allowed is to verify if the two considered process parameters influence significantly the extracted features.

Finally, regression analysis is allowed to write the analytical relationship among the force feature X_{rms}, that feels the most significant effects of the mill wear, the diamond mill wear (T), and the considered process parameters (u and p).

Results

Discussion

ANOVA analysis proved that both the diamond mill wear and the two considered process parameters, feed speed and depth of cut, influence significantly all the 6 considered features of F_x and F_y signals. The interactions between the three factors, mill wear, feed speed, and depth of cut, do not significantly influence anyone of the 6 features of F_x and F_y signals. Figure 7 shows the average values of the 6 features extracted from F_x signal for each value of the considered process parameters, feed speed, depth of cut, and mill wear. We can see that the mean value X_{avg} and X_{rms} value of F_x signal strongly increase with the increase of the feed speed (u), of the depth of cut (p), and of the mill wear (T). However, the three parameters sigma, skewness (S), and crest factor (CF) seem to be less affected by the increase of the process parameters, since their trend is slowly. Finally, the influence of the process parameters on the kurtosis (K) is not clear, since we cannot identify a univocal trend. The same effects were observed for F_y signal.

ANOVA analysis demonstrated that X_{rms} is the feature that feels the most significant effects of diamond mill wear for the three considered force signals. Therefore, X_{rms} was considered to deepen the analysis in order to monitor mill wear by acquiring cutting force signal during stone cutting. The other features were neglected in the following stages of analysis.

Plots of the X_{rms} feature versus diamond mill wear were produced together with cutting force relationship with feed rate and depth of cut (see Figures 8 and 9). The x-axis of Figures 8 and 9 shows the 5 stages of diamond mill wear distributed along mill tool life, as introduced in Section 3. 1. The 5 stages of wear were put into relationship with the protrusion height of the diamond grits by the bond observed on mill surface, as shown in Figure 10.

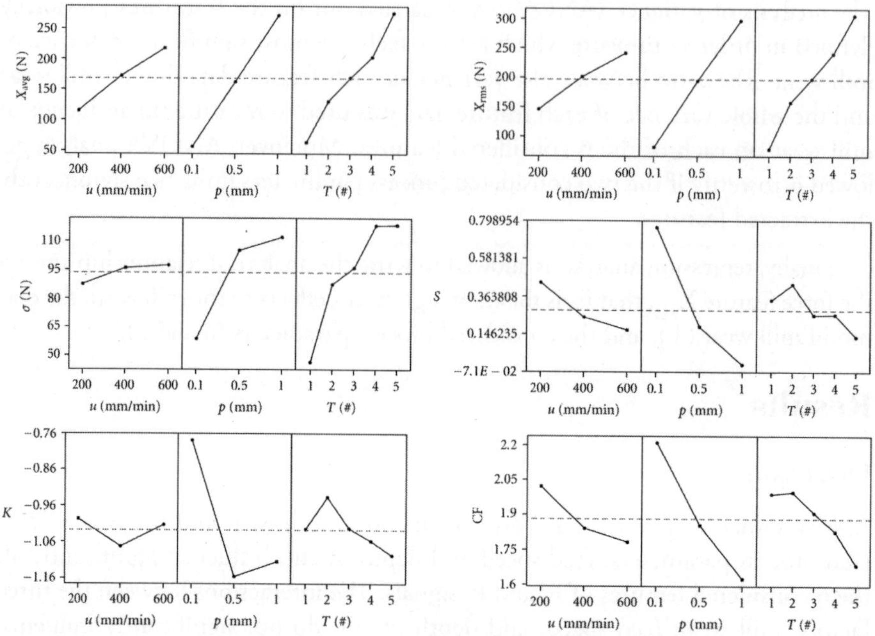

Figure 7: Main effect plots of features extracted by Fx signal versus feed speed (u), depth of cut (p), and mill wear (T).

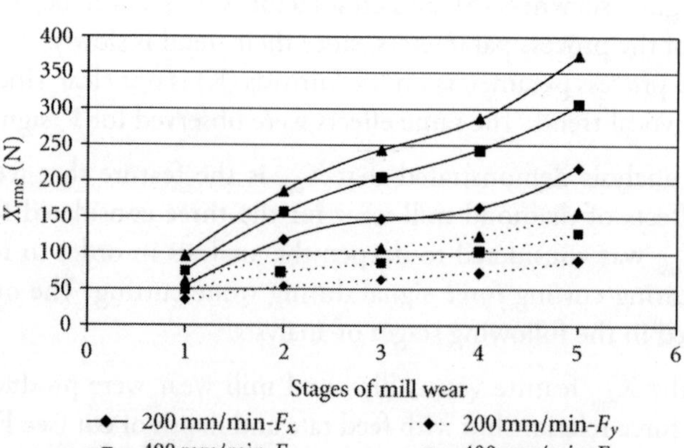

Figure 8: X_{rms} feature of Fx and Fy versus diamond mill wear for the three values of feed speed (200, 400, and 600 mm/min).

Inspection of Figures 8 and 9 shows that the x-component of the cutting force is higher than the y-component, since the angle of contact between mill and stone is lower than 45°, due to the low value of the depth of cut. Moreover, X_{rms} increases significantly with the mill wear for the same value of the feed rate or of the depth of cut, X_{rms} increases strongly with the feed speed or with the depth of cut for the value of the diamond mill wear too.

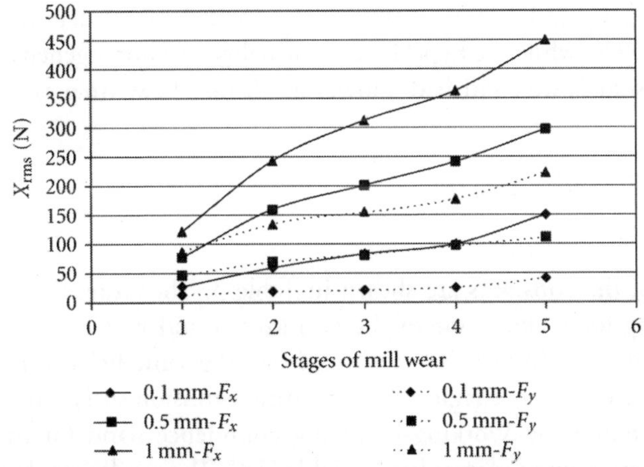

Figure 9: X_{rms} feature of Fx and Fy versus diamond mill wear for the three values of depth of cut (0. 1, 0. 5, and 1 mm).

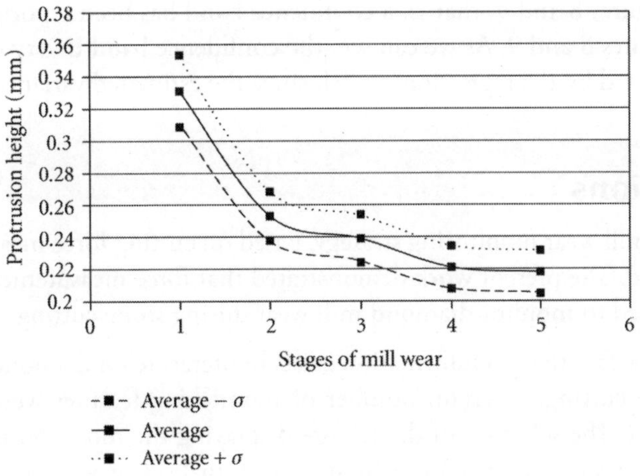

Figure 10: Height of the diamond grits versus stages of mill wear.

Regression analysis underlined that the analytical relationship among X_{rms} feature, mill wear T, feed speed va, and depth of cut p is given by the following expression for the two considered signals, F_x and F_y, respectively,

$$\log(X_{rms})=k_1+k_2 \cdot \log(T)+k_3 \cdot \log(v_a)+k_4 \cdot \log(p), \quad (7)$$

where k_1, k_2, k_3, and k_4 are constants. The coefficients of determination were higher than 98%, while the hypotheses (normality and homogeneity of variance) about the residuals were satisfied. However, (7) may be written as

$$X_{rms} = e^{k_1} \cdot T^{k_2} \cdot v_a^{k_3} \cdot p^{k_4}. \quad (8)$$

The values of the constants are shown in Table 3. Therefore, the measurement of the cutting force along one of the two axes, x and y, is enough to monitor diamond mill wear. In fact, X_{rms} feature shows the same behaviour for the two axes. The 95% confidence region for the entire regression surface has been evaluated by extending the Working-Hotelling confidence band for the regression line when there is one independent variable [18]. The results are shown in Table 4 where we have reported the limits of the confidence band together with the experimental value and the corresponding value given by the regression model (all the values are in Newton). The results in Table 4 match with the projection curves of Figures 8 and 9, that is, a confidence band has been reported for each curve of Figures 8 and 9. As we can see, the confidence band is very close to the values estimated by the regression model, since the estimated variance has a very small value.

Conclusions

A diamond mill wear monitoring strategy, based on cutting force measurements, was presented. The present work demonstrated that force measurements may be effectively used to monitor diamond mill wear during stone cutting.

Due to the fact that no information exists in literature on diamond mill wear during stone cutting, a certain number of monitoring features were extracted from the data. The selection of the feature displaying the most consistent trend towards mill wear was carried out by means of analysis of variance (ANOVA) results. The selected feature Xrms was put into relationship with mill wear and process parameters.

Table 3: Constants of regression equations.

	Regression constants			
Features	k_1	k_2	k_3	k_4
X_{rms} of F_x	2.03	0.464	0.578	0.851
X_{rms} of F_y	1.64	0.475	0.846	0.521

Table 4: 95% confidence band for regression surface (the values are in N).

Feed speed			200 mm/min					400 mm/min					600 mm/min		
Tool wear	1	2	3	4	5	1	2	3	4	5	1	2	3	4	5
Upper confidence limit	57,3	110,7	145,5	199,5	224,6	84,7	142,1	200,5	256,1	332,5	102,1	171,4	242,0	309,0	401,1
Experimental value	54,7	128,5	147,2	187,3	216,1	81,4	157,6	205,3	242,4	322,2	102,4	186,0	243,3	287,1	392,8
Model value	57,0	110,4	145,2	199,2	224,3	84,4	141,8	200,3	255,8	332,2	101,9	171,1	241,7	308,8	400,9
Lower confidence limit	56,7	110,2	145,0	199,0	224,1	84,1	141,6	200,0	255,6	331,9	101,6	170,9	241,4	308,5	400,6
Depth of cut			0.1 mm					0.5 mm					1 mm		
Tool wear	1	2	3	4	5	1	2	3	4	5	1	2	3	4	5
Upper confidence limit	30,6	58,9	80,5	106,0	119,6	80,1	144,3	203,6	260,1	314,4	119,4	215,3	295,0	388,1	469,2
Experimental value	25,9	62,2	82,9	103,8	135,1	76,8	159,4	201,0	241,2	296,8	121,2	243,0	302,8	362,5	449,8
Model value	30,3	58,6	80,2	105,8	119,3	79,8	144,0	203,4	259,8	314,2	119,2	215,0	294,7	387,8	469,0
Lower confidence limit	30,1	58,4	79,9	105,5	119,1	79,6	143,7	203,1	259,5	313,9	118,9	214,7	294,4	387,6	468,7

The X_{rms} feature of cutting force signals along x and y axes is significantly influenced by mill wear, feed speed, and depth of cut, while the interactions seem not to significantly influence X_{rms} feature. The cutting force signal along z-axis is very near to 0, while the cutting force signal measured along x-axis is higher than that of y-axis.

The relationship among X_{rms} and mill wear, feed speed, and depth of cut shows an exponential trend for both the considered signals, F_x and F_y. The identified relationship may be used to predict diamond mill wear by measuring cutting force along one of the axes, x or y.

Finally, the present work suggests that even under varying cutting conditions, involving different machining parameters, it is possible to control, in a reliable and accurate way, diamond mill wear during stone cutting by measuring the cutting force along only one axis, x or y, respectively, since the corresponding signals show the same trends.

Acknowledgments

This work has been carried out with the funds received by the Italian Ministry of Instruction, University and Research (M. I. U. R.) within the project entitled "Intelligent sensor monitoring of stone machining" (MIUR-PRIN 2002).

References

1. D. E. Dimla, Snr. , "Sensor signals for tool-wear monitoring in metal cutting operations—a review of methods, " International Journal of Machine Tools and Manufacture, vol. 40, no. 8, pp. 1073–1098, 2000.
2. H. V. Ravindra, Y. G. Srinivasa, and R. Krishnamurthy, "Acoustic emission for tool condition monitoring in metal cutting, " Wear, vol. 212, no. 1, pp. 78–84, 1997.
3. Y. M. Ertekin, Y. Kwon, and T. -L. Tseng, "Identification of common sensory features for the control of CNC milling operations under varying cutting conditions, " International Journal of Machine Tools and Manufacture, vol. 43, no. 9, pp. 897–904, 2003.
4. D. E. Dimla, Snr. and P. M. Lister, "On-line metal cutting tool condition monitoring. I: force and vibration analyses, " International Journal of Machine Tools and Manufacture, vol. 40, no. 5, pp. 739–768, 2000.
5. A. K. Ghani, I. A. Choudhury, and Husni, "Study of tool life, surface roughness and vibration in machining nodular cast iron with ceramic tool, " Journal of Materials Processing Technology, vol. 127, no. 1, pp. 17–22, 2002.
6. T. I. El-Wardany, D. Gao, and M. A. Elbestawi, "Tool condition monitoring in drilling using vibration signature analysis, " International Journal of Machine Tools and Manufacture, vol. 36, no. 6, pp. 687–711, 1996.
7. G. H. Lim, "Tool-wear monitoring in machine turning, " Journal of Materials Processing Technology, vol. 51, no. 1–4, pp. 25–36, 1995.
8. J. C. Chen and W. -L. Chen, "A tool breakage detection system using an accelerometer sensor, " Journal of Intelligent Manufacturing, vol. 10, no. 2, pp. 187–197, 1999.
9. N. K. Mehta, P. C. Pandey, and G. Chakravarti, "An investigation of tool wear and the vibration spectrum in milling, " Wear, vol. 91, no. 2, pp. 219–234, 1983.
10. D. Guinea, A. Ruiz, and L. J. Barrios, "Multi-sensor integration—an automatic feature selection and state identification methodology for tool wear estimation, " Computers in Industry, vol. 17, no. 2-3, pp. 121–130, 1991.
11. R. G. Silva, R. L. Reuben, K. J. Baker, and S. J. Wilcox, "Tool wear monitoring of turning operations by neural network and expert system classification of a feature set generated from multiple sensors, "

Mechanical Systems and Signal Processing, vol. 12, no. 2, pp. 319–332, 1998.

12. C. Scheffer and P. S. Heyns, "Wear monitoring in turning operations using vibration and strain measurements, " Mechanical Systems and Signal Processing, vol. 15, no. 6, pp. 1185–1202, 2001.
13. S. Kalpakian and S. R. Schmid, Manufacturing Engineering and Technology, Prentice-Hall, Upper Saddle River, NJ, USA, 2001.
14. S. Jahanmir, M. Ramulu, and P. Koshy, Machining of Ceramics and Composites, Marcel Dekker, New York, NY, USA, 1999.
15. D. N. Wright, H. Wapler, and H. K. Tönshoff, "Investigations and prediction of diamond wear when sawing, " CIRP Annals: Manufacturing Technology, vol. 35, no. 1, pp. 239–244, 1986.
16. W. Polini and S. Turchetta, "Test protocol for micro-geometric wear of sintered diamond tools, " Wear, vol. 257, no. 3-4, pp. 246–256, 2004.
17. S. Turchetta, W. Polini, and L. Carrino, "Wear progression of diamond mills: a micro-geometric study, " in Proceedings of the European Conference on Hard Materials & Diamond Tooling (Euro PM '02), Lausanne, Switzerland, October 2002.
18. J. Neter and W. Wasserman, Applied Linear Statistical Models, R. D. Irwin, Homewood, Ill, USA, 1974.

Measurement and Simulation of Pollutant Emissions from Marine Diesel Combustion Engine and Their Reduction by Ammonia Injection

Nader Larbi and Jamel Bessrour

ABSTRACT

Taking into account the complexity and cost of a direct experimental approach, the recourse to a tool of simulation, which can also predict inaccessible information by measurement, offers an effective and fast alternative to apprehend the problem of pollutant emissions from internal combustion engines. An analytical model based on detailed chemical kinetics employed to calculate the pollutant emissions of a marine diesel engine gave satisfactory results, in general, compared to experimentally measured results. Especially the NO emission values are found to be higher than the limiting values tolerated by the International Maritime Organization (IMO). Thus, this study is undertaken in order to reduce these emissions to the maximum level. The reduction of pollutant emissions is apprehended with ammonia injection.

Introduction

The energies produced by the diesel engines of strong power are largely used in marine propulsion because of their favorable reliability and their significant output. However, the increasingly constraining legislations, aimed at limiting the pollutant emissions from the exhaust gas produced by these engines, tend to call into question their supremacy [1, 2].

Most bibliographical studies on the various approaches to estimate the rate of emissions cover many established models such as

(i) the direct models that lead to digital resolution of Navier Stokes, energy conservation, and phenomenological species equations that have given rise to codes: CONCHAS, KIVA, and KIVA II,

(ii) the phenomenological models that simulate combustion, using formulations based on different physical and chemical phenomena that were experimentally observed and described,

(iii) the empirical models establishing kinetics patterns reduced to a dozen variables necessary for calculations of turbulent flames equations which lead to calculate the formation of pollutants.

Many other studies were undertaken, on the experimental level as well as on the analytical level, in order to study the mechanisms which govern the formation of the various produced pollutants. The study of combustion chemistry enjoys an undeniable anticipation, and has now acquired a certain maturity [3–5].

The analysis of the pollutant emissions and their reduction in the exhaust gas of the semirapid turbocharged marine diesel engine constitutes the principal objective of this study.

The various simulations carried out in this work are based on the computer code marketed by Reaction Design, the CHEMKIN package, which is developed by the Sandia National Laboratories.

Comparisons of the numerical predictions with the experimental results carried out on a real unit at use aboard a car ferry ship made it possible to analyze the validity of the numerical results.

Experimental Study

The unit selected for measurements is a power generating unit in use on a car ferry ship during its docking in a harbor. The diesel engine unit is a WARTSILA NSD type 6R32 LNE having characteristics given in Table 1 with a nominal power of 2460 kW. The measured values taken in exhaust gas were carried out using a Testo350 (Precision +/−5% mV) gas analyzer, which made it possible to post with precision and in real-time the percentage by volume of oxygen (O^2), percentage by volume of carbon dioxide (CO^2), nitric oxide (NO), sulfur dioxide (SO^2), temperature, and pressure.

Table 1: Characteristics of the diesel engine of the power generating unit.

The measurements were carried out for various powers: 640, 900, 1370, and 1470 kW.

To determine the emitted quantities in grams per cycle per cylinder (g/cyc/cyl) and in (g/kWh), it is necessary to deduce the volume (Vm) of the exhaust gas at the measurement point at atmospheric pressure (Pa) and for exhaust gas temperature (Tm)

$$V_m = \frac{mtR_gT_m}{P_a}, \quad (1)$$

where the volume is calculated from the ideal gas equation of state. The mass conservation law enables us to deduce the mass with the admission condition

$$m_t = m_{air} + m_{comb},$$

where it is assumed

$$m_{air} = 15\lambda_a m_{comb}. \quad (2)$$

The volume of exhaust gas at the measurement point is then

$$Vm = Vcyl \frac{T_m}{T_{adm}} \frac{P_{adm}}{P_a}\left(1 + \frac{1}{15\lambda_a}\right). \quad (3)$$

Figure 1 presents the preliminary experimental results obtained on the chosen unit.

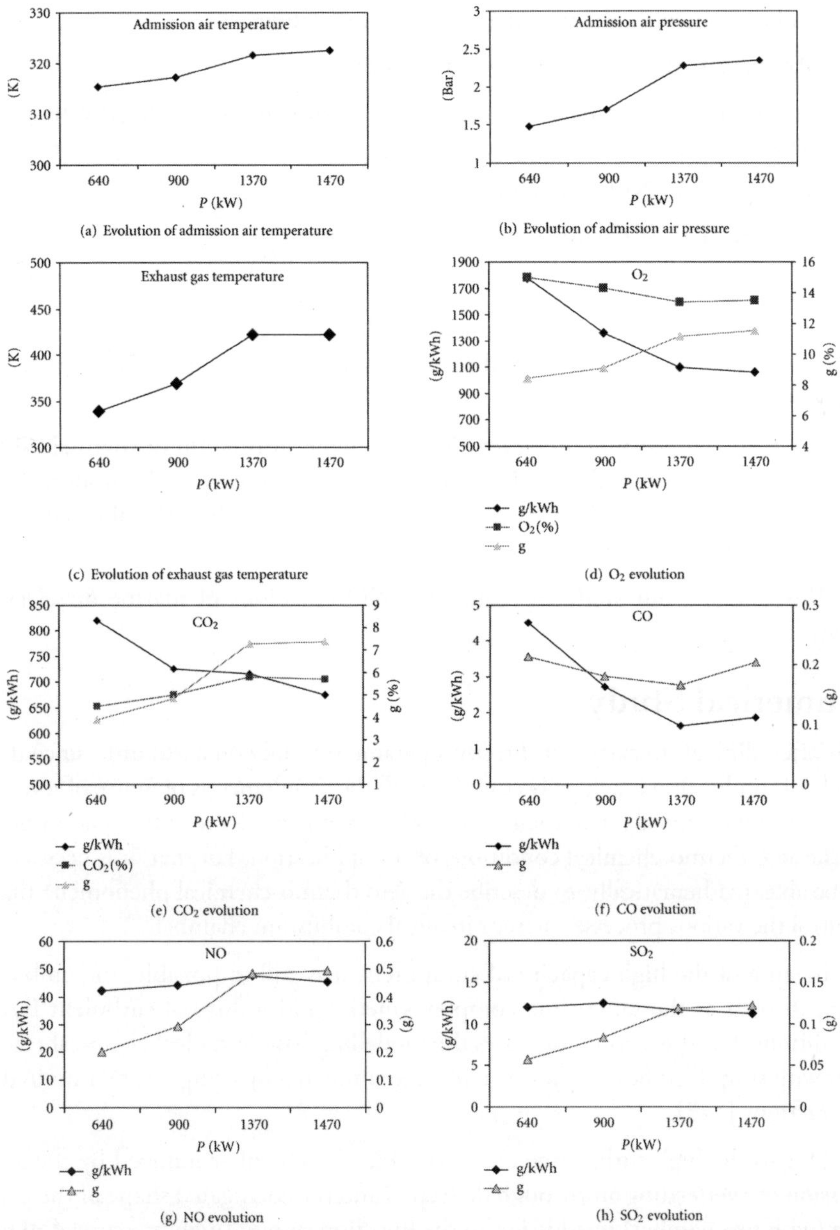

Figure 1: Parameters measured with gas analyzers.

As a consequence of the increase in power we note the following.

(i) A greater quantity of exhaust gas increasing the mode of the turbocharger and consequently the pressure of air of overfeeding.

(ii) An increase in overfeeding air temperature.

(iii) A decrease in percentage by volume of O2 with an increase in percentage by volume of CO2.

(iv) An increase in NO emission.

(v) An increase in SO2 emission.

According to these notes, we can conclude that there is an improvement of combustion for the passage from the low power to the semimaximum power of the generating unit.

The cycle temperature increases causing a higher formation of thermal NO. In addition, the admitted quantity of air is larger and consequently supports the formation of NO in greater quantity. This last is higher than the allowed NO emission by the IMO.

This initiated our study to reduce the NO emissions of marine diesel engines.

Numerical Study

It is often difficult to carry out directly a parametric study on a real unit, since the study of combustion is generally complex with the simultaneous presence of several physical and chemical phenomena that vary from one model to another according to the aero thermo-chemical conditions of the application. For that, it is necessary to be able, mathematically, to describe the aero thermo-chemical phenomena that control the various processes present in a real combustion chamber.

In spite of the high capacity of computers, it is still impossible, in combustion, to treat at the same time complex kinetics and industrial turbulent flow conditions. For this reason, one uses the modelling based on ideal chemical reactors with simplified flows. These reactors are simulated by using several hundreds of reactions [5–7].

Due to the high turbulence in the combustion chamber induced by the admission of overfeeding air through deflectors and the corrugated shape of the piston (with prechamber) in addition to the injection system (high pressure: 450 to 600 Bar) [8, 9] that allows a good distribution of the fuel jet in the combustion chamber [10–13], one can admit that the reagents mixture is homogeneous, and we assume that the combustion process can be modelled by a perfectly stirred reactor (PSR) [14, 15].

In a PSR, the mixing in the reactor chamber is supposed to be intense and, thus, it is assumed that the temperature and composition in the reactor are uniform through the reactor volume.

A description of the process occurring within the PSR is obtained by relating the conservation of mass and energy to the generation of chemical species within the reactor volume [5, 16].

(i) The species conservation equation is

$$m\,(Y_k - Y_k^*) - \omega_k M_k V = 0,\ k=1, \ldots, K. \quad (4)$$

(ii) The energy conservation equation is

$$m\sum_{k=1}^{K}\left(Y_k h_k - Y_k^* h_k^*\right) + Q = 0. \quad (5)$$

(iii) The nominal residence time is

$$\tau = \frac{pV}{m}, \quad (6)$$

where the mass density ρ is calculated from the ideal gas equation of state

$$\rho = \frac{P\overline{M}}{R_g T}. \quad (7)$$

From this set of (K+1) nonlinear algebraic equations, solutions for the temperature and mass fractions are obtained. Even though one seeks the solution to the steady-state equations, the computational algorithm often requires a partial solution of the related transient problem.

The analogous time-dependent equations for mass conservation of each species are

$$\frac{dY_k}{dt} = \frac{-\left(Y_k - Y_k^*\right)}{\tau} + \frac{\omega M_k}{\rho}, \quad (8)$$

and the time-dependent energy conservation equation is

$$Cp\frac{dT}{dt}=\frac{1}{\tau}\sum_{k=1}^{K}Y_k^*(h_k^*-h_k)-\sum_{k=1}^{K}\left(\frac{h_k M_k \omega_k}{\rho}\right)-\frac{Q}{\rho V}. \quad (9)$$

The net chemical production rate ωk. of each species results from a competition between all the chemical reactions involving that species. Each reaction proceeds according to the law of mass action and the forward rate coefficients (kf) are in modified Arrhenius form:

$$k_f = AT^{\beta}\exp\left(\frac{-E_A}{R_g T}\right). \quad (10)$$

The simulation of the internal combustion with high-turbulent conditions of an engine is carried out by the CHEMKIN code using 450 elementary reaction mechanisms between 77 species including sulfur (case of the fuel oil). Specifically, the internal combustion engine model (ICEM) has been used to simulate the temporal behavior of the engine combustion [16].

The process of ignition of fuel used by the code is governed by its temperature of autoignition.

The relation of volume swept by the piston divided by clearance volume is as follows:

$$\frac{V_{cyl}}{V_0}=1+\frac{C-1}{2}\left[G+1-\cos\alpha-\sqrt{G^2-\sin^2\alpha}\right]. \quad (11)$$

The convective heat transfer coefficient between the gas and cylinder wall obtained from the generalized heat transfer correlation in terms of a Nusselt number is

Nuh=a RebPrc, Nuh≡hSλ. (12)

The heat loss is calculated at each step in time according to

$Q_{wall}=hS(T-T_{wall})$. (13)

The Woschni correlation [16] allows a more accurate estimation of the average cylinder gas speed used in the definition of the Reynolds number for the heat-transfer correlation.

The velocity used in the Reynolds number definition in (14) is an estimation of the average cylinder gas velocity, Z, instead of the mean piston speed

$$Re \equiv \frac{D\bar{Z}\rho}{\mu}. \quad (14)$$

To obtain the average cylinder gas velocity, Woschni proposed a correlation that relates the gas velocity to the mean piston speed and to the pressure rise due to combustion:

$$Z = \left[C_{11} + C_{12} \frac{V_{swirl}}{\bar{S}_p} \right] \bar{S}_p + C_2 \frac{V_d T_i}{P_i V_i} (P - P_{motored}). \quad (15)$$

The chosen composition of species weight of the fuel introduced in the code is composed of 86. 6% in mass of carbon, 10. 9% in mass of hydrogen, and 2. 5% in mass of sulfur.

The numerical results are established according to pressure, which varies from 1. 7 to 2. 35 Bar, an overfeeding air temperature, which varies from 317 to 322 K, and an equivalence ratio, which varies from 0. 6 to 0. 8, corresponding to the values measured with the powers which vary from 900 to 1470 kW and for an ambient temperature of 306 K.

The first objective of the study was to analyze the influence of the equivalence ratio on the course of the combustion process and on the pollutant emissions related to the real engine conditions (overfeeding air temperature, overfeeding air pressure, engine speed, etc.).

The numerical conversion of the results molar fraction (MF) into grams (m_k) (16), for each species k (k varying from 1 to 77), is obtained while passing by the molar masses M_k and the total mass

$$m_t = \sum_{k=1}^{77} m_k, \; m_k = m_t \frac{M_k MF_k}{\sum_{k=1}^{77} M_k MF_k} \text{ in grams. } (16)$$

Figure 2 illustrates the influence of the equivalence ratio on the progress of the combustion and the pollutant emissions according to the crank rotation an-

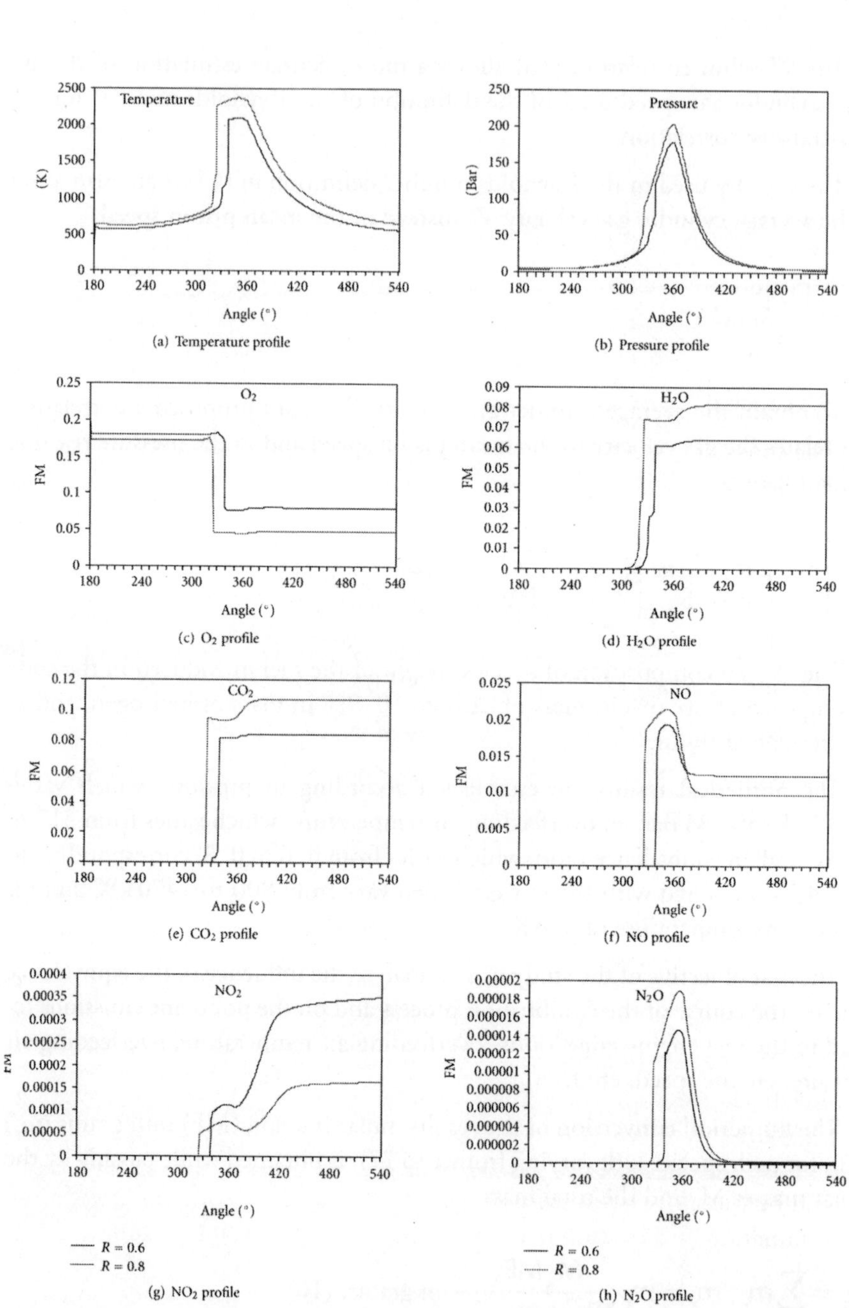

Figure 2: Evolutions of the temperature, the pressure, and the pollutant emissions in the combustion chamber.

gle for the same running condition. The observed jump of the parameters corresponds to the beginning of the combustion which starts, with an advance, with regard to the piston top dead center (360°) and with a delay of ignition which varies with the equivalence ratio.

For a reduction in the equivalence ratio (from R=0. 8to R=0. 6), we note that the excess of total air leads to a dilution of exhaust gas which decreases its temperature and pressure, an increase in the ignition delay, a decrease in the total time of combustion, a considerable increase in NO as of NO2 and SO3 emissions, a considerable decrease in H2O and SO2 CO2 emissions.

We conclude that there is an improvement in the combustion process.

The considerable increase in the excess of air and consequently that of atmospheric nitrogen support the formation of NOx in quantity to the exhaust gas which increases.

For a poor mixture, the fuel jet of a smaller quantity of fuel undergoes a faster oxidation, which has, as a consequence, an increase in the time of ignition.

The different values expressed in g/cycle/cylinder and g/kWh are given by considering a low calorific value of 42. 000 kJ/Kg and a total output of 0. 4.

The content of CO_x is made up mainly of 99. 9% to CO2.

The content of NO_x is made up mainly of 97. 5% in NO and 2. 5% in NO2 in the poor mixture; consequently, the NOx emission undergoes an increase since it follows the same evolution of the majority species which is the NO.

The content of SO_x is made up mainly of 95% of SO2 and 5% of SO3 in the poor mixture; consequently, the emission of SOx undergoes a small decrease.

Figures 3, 4, and 5 show comparisons between computed and measured CO2-, NO-, and SO2-emitted quantities in the exhaust gas as a function of engine power.

Relatively good agreements are observed especially for emission expressed in g/kWh. One can conclude that the model of the adopted calculation allows qualitative and quantitative results, that are, in general, satisfactory. But it should be well noted that the precision of the results depends on the adequate determination of the precision of measures and the assumption adopted in the model.

Figure 6 shows the IMO [17] curve for the limiting NO_x emissions according to the nominal speed of the engine. For a speed of 750 rpm, the NOx is limited to a value of 12. 2 g/kWh.

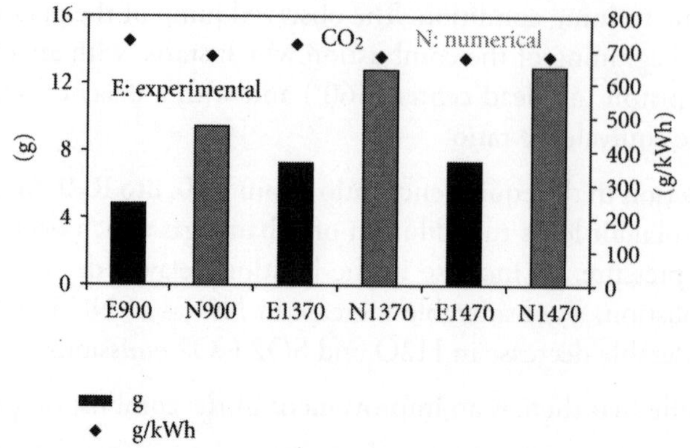

Figure 3: Evolution of numerical and experimental results for the CO2 emission.

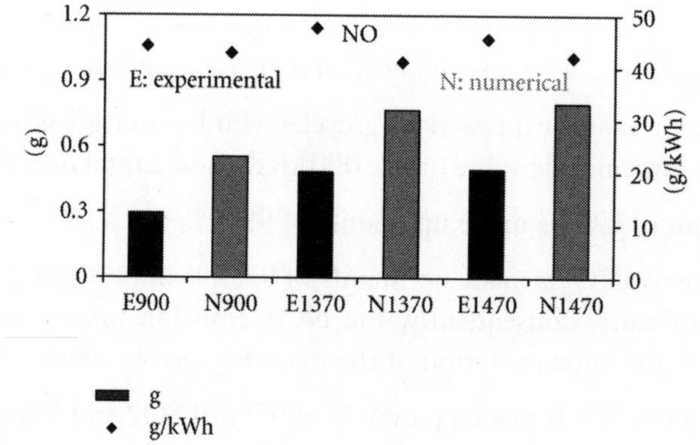

Figure 4: Evolution of numerical and experimental results for the NO emission.

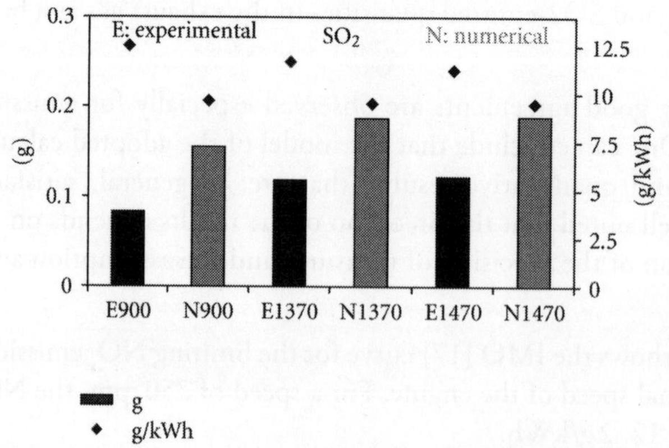

Figure 5: Evolution of numerical and experimental results for the SO2 emission.

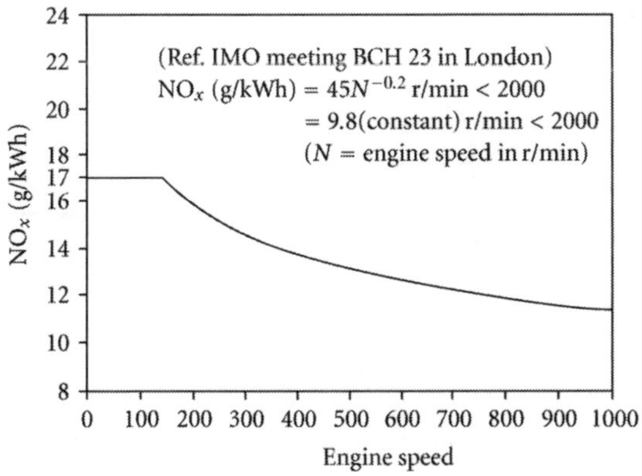

Figure 6: Curve IMO limiting the NO_x emissions according to the nominal speed of the engines [14].

All the values of NO (g/kWh) in Figure 1(g), and for the various powers, are higher than the limiting value tolerated by the IMO.

Study of Ammonia Injection

Therefore, a subsequent study has been done to reduce this NO emission [18–21] by ammonia injection [22]. The objective of this study is to analyze the influence of ammonia injection in the combustion chamber of the diesel engine on the different pollutant emissions.

The study of influence of ammonia injection is that corresponding to the measured maximum power (1470 kW) having a temperature of admission air of Ta=322 K and a pressure of admission air of Pa=2. 35 Bar.

The numerical results are established according to NH3 mass in reagent which varies from 0 to 2. 18 g, and for air mass which varies from 74. 38 to 70. 7 g.

The volume of NH3 mixed with the admission air is

$$V_{NH3} = \frac{XV_{cyl}}{100},\qquad V_{air} = \frac{(100 - X)V_{cyl}}{100}, \tag{17}$$

where X is the percentage in volume of NH3 mixed with admission air.

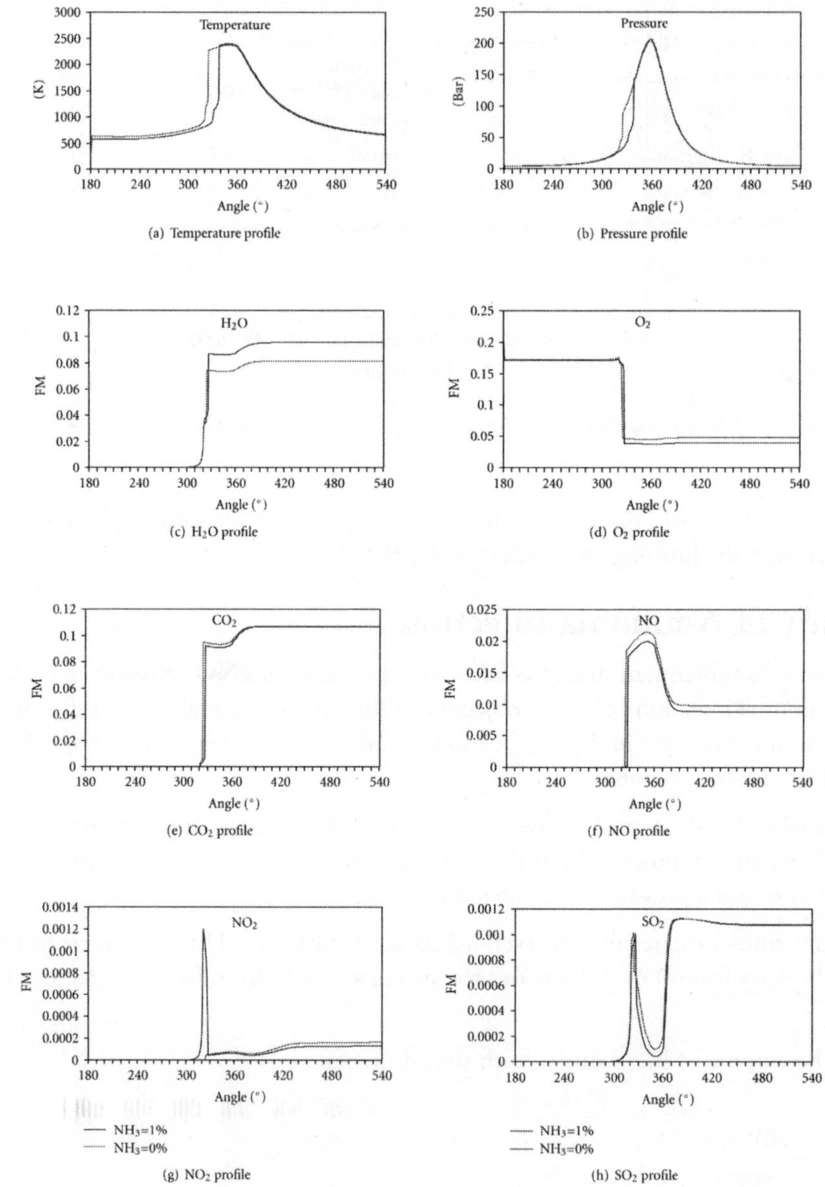

Figure 7: Evolutions of the temperature, the pressure, and the pollutant emissions.

Figure 7 illustrates the influence of ammonia injection on the course of combustion and the pollutant emissions. It represents the variations of the pressure, the temperature, and the pollutant emissions in the cylinder as a function of the crank rotation angle with and without the ammonia injection. A substantial reduction in NO and NO2 emissions is noticed, while the CO2 and SO2 emissions remain constant.

Figure 8 illustrates the influence of ammonia injection on the average combustion pollutants emissions expressed in g/cyc/cyl and g/kWh. A considerable decrease in NO and NO2 emissions is now way down and meets the IMO index requirement. A considerable increase in H2O quantity is observed.

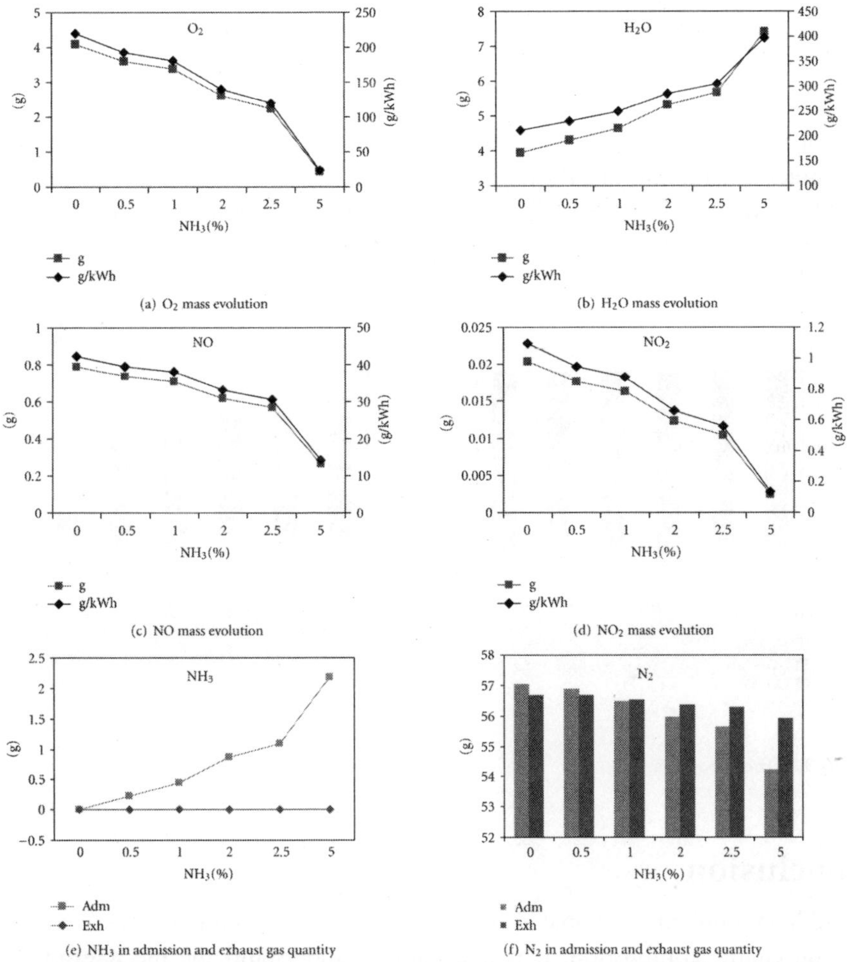

Figure 8: Evolutions of pollutant emissions.

The N2 quantity in admission air undergoes a decrease and the N2 quantity in exhaust gas remains constant with a percentage in volume of NH3 which varies from 0 to 0. 5. For the other percentage, we note a considerable increase in N2 between admissions and exhaust gas quantity.

We can conclude that in combustion process, the NO and NO2 were transformed to H2O and N2 by these two following reactions:

4NO+4NH3+O2→4N2+6H2O,

6NO2+8NH3→7N2+12H2O.

For the average of 1% in volume of ammonia mixed with the admission air, one notes a reduction of NO on average of 11. 7% g/kWh and a reduction of NO2 on average of 21% g/kWh, consequently, the NO_x emission (Figure 9) undergoes a reduction on average of 12% g/kWh.

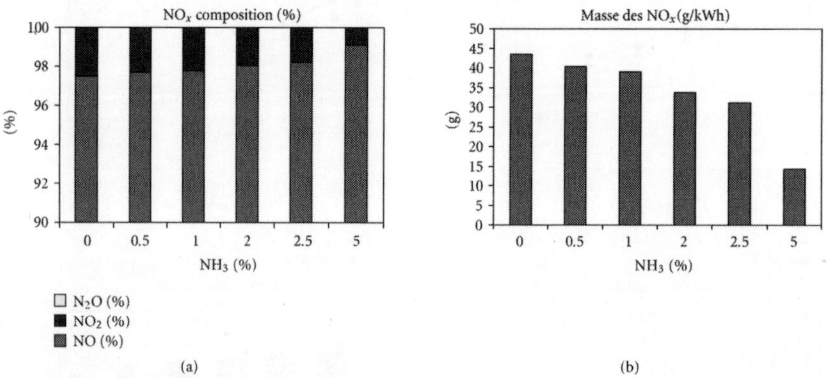

Figure 9:NOx composition.

Conclusion

To address problems of global air pollution due to the pollutant emission from fuel oil engine combustion, it is necessary to understand the mechanisms by which pollutants are produced in combustion processes. In the present work, an experimental and numerical study is carried out on a unit of real use aboard a car ferry ship. A numerical model based on a detailed chemical kinetics scheme is used to calculate the emissions of CO2, NO, and SO2 in an internal combustion engine model for the same characteristics of the real unit. In general, the experimental and numerical results featured good agreements especially at high power and for excess air. For the study of the reduction of pollutant emissions of the diesel engine by the use of ammonia injection, substantial decrease in NO was observed which made the unit meet the IMO regulations.

Nomenclature

A:	Pre-exponential factor
a, b, and c:	Constants
C:	Modeling parameters
C_p:	Thermal capacity (J · Kg–1 · K–1)
D:	The engine bore diameter for heat transfer(cm)
E_A:	Activate energy (J)
G:	Ratio of connecting rod to crank-arm radius
h:	Heat transfer coefficient (w · m–2 · K–1)
h_k:	Specific enthalpy of the kth species(J/Kg)
k_f:	Forward rate coefficient
m:	Mass flow rate (Kg/s)
MF:	Molar fraction
MF_k:	Molar fraction of the kth species
M_k:	Molar mass of the kth species(g/mol)
m_t:	Total mass of the reactants(g)
m_{comb}:	Mass of the fuel (g)
m_{air}:	Mass of the air(g)
m_k:	Mass of the kth species (g)
Nu_h:	Nusselt number
P_a:	Atmospheric pressure (Bar)
P_{adm}:	Admission air pressure (Bar)
P_i:	Initial pressure inside the cylinder
$P_{motored}$:	The motored cylinder pressure
Pr:	Prandtl number.

Q:	Reactor heat (cal · s−1)
Q_{wall}:	Wall reactor heat loss (cal · s−1)
R:	Equivalence ratio
R_g:	Universal gas constant (J · kg−1 · K−1)
Re:	Reynolds number
S:	The surface area for heat transfer (cm2)
$\bar{S}_p$:	Mean piston speed (m/s)
T:	Temperature (K)
T_{adm}:	Admission air temperature(K)
T_i:	Initial temperature inside the cylinder
T_m:	Exhaust gas temperatureat the measurement point(K)
T_{wall}:	Chamber wall temperature(K)
V:	Reactor volume (m3)
V_o:	Clearance volume (m3)
V_{cyl}:	Volume swept by the piston(m3)
V_d:	Displacement volume
V_i:	The initial volume inside the cylinder
V_m:	Exhaust gas volume at the measurement point (m3)
V_{NH3}:	Volume of ammonia
V_{air}:	Volume of admission air
V_{swirl}:	Swirl velocity
X:	Percentage in volume of NH3 mixed with admission air
Y_k:	Mass fraction of the kth species
Z:	Gas velocity
λ:	Gas conductivity (w · m−1 · K−1)
λ_a:	Air excess
α:	Crank angle (°)
β:	Temperature exponent in the rate coefficient
ρ:	Density (Kg/m3)
τ:	Residence time in the reactor (ms)

ω_k:	Molar rate of production of the kth species
*:	Inlet condition
IMO:	International Maritime Organisation
ICEM:	Internal combustion model engine
PSR:	Perfectly stirred reactor.

References

1. ANPE, PAM, and FEM, "Atelier sur la lutte contre la pollution due à des activités menées à terre, " Tunisie, 2005.
2. X. Tauzia, Simulation de l'influence des paramètres de fonctionnement des moteurs Diesel suralimentés semi-rapides sur les émissions polluantes, Ph. D. thesis, Ecole Doctorale Sciences pour l'ingénieur de Nantes, Nantes, France, 1998.
3. R. Schaal, La Cinétique Chimique Homogène, Presses Universitaires deFrance, Paris, France, 1971.
4. C. T. Bowman, "Kinetics of pollutant formation and destruction in combustion, " Progress in Energy and Combustion Science, vol. 1, no. 1, pp. 33–45, 1975.
5. H. Mohamed, H. B. Ticha, and S. Mohamed, "Simulation of pollutant emissions from a gas-turbine combustor, " Combustion Science and Technology, vol. 176, no. 5-6, pp. 819–834, 2004.
6. P. Gateau and P. Franco, "Application des modèles mécanistiques de cinétique chimique aux combustions industrielles, " Revue de l'Institute Français du Pétrole, vol. 46, no. 3, pp. 407–419, 1991.
7. R. J. Kee, M. E. Coltrin, and P. Glarborg, Chemically Reacting Flow: Theory and Practice, John Wiley & Sons, Hoboken, NJ, USA, 2003.
8. I. Baz, Contribution à la caractérisation de la cavitation dans les injecteurs Diesel à haute pression, Ph. D. thesis, École Centrale de Lyon, Lyon, France, 2003.
9. H. Takahashi, H. Yanagisawa, S. Shiga, T. Karasawa, and H. Nakamura, "Analysis of high-pressure diesel spray formation in the early stage of injection, " Automization and Sprays, vol. 7, no. 1, pp. 33–42, 1997.
10. C. Jean, "La mécanique des Moteur Diesels marins, " Institut maritime du Québec, Cégep de Rimouski, Quebec, Canada, 1999.
11. A. Haupais, "Combustion dans les moteurs Diesel, " Techniques de l'ingénieur, Traité mécanique et chaleur, 1992.

12. J. Surugue and M. Barrere, Combustion, Librairie polytechnique Béranger, Département Technique des presses de la cité, Paris, France, 1963.

13. R. Borghi, M. Destriau, and G. De Soete, La Combustion et les Flammes, Editions Technip, Paris, France, 1995.

14. P. Glarborg, R. J. Kee, J. F. Grcar, and J. A. Miller, "PSR: a fortran program for modeling well-stirred reactor, " Sandia National Laboratories, Livermore, Calif, USA, 1986.

15. R. J. Kee, J. Warnatz, J. A. Miller, and A. Fortran, "Computer code package for the evaluation of gas phase viscosities, heat conductivities, and diffusion coefficients, " Sandia National Laboratories, Livermore, Calif, USA, 1983.

16. Reaction Design, "Input Manual, Theory Manual, " Chemkin Collection RD01402, 2005.

17. Annexe VI à MARPOL 73/78, "Régles Relatives à la prevention de la pollution de l'atmosphère par les navires et Code technique sur le contrôle des émissions d'oxydes d'azote provenant des moteurs Diesel marins, " OMI: Londres, 1998.

18. N. H. Abu-Hamdeh, "Effect of cooling the recirculated exhaust gases on diesel engine emissions, " Energy Conversion and Management, vol. 44, no. 19, pp. 3113–3124, 2003.

19. A. Velji, W. Remmels, and R. M. Schmidt, "Water to reduce NOx Emissions in Diesel engines a basic study, " in Proceedings of 21st CIMAC Congress Interlaken (CIMAC '95), London, UK, 1995.

20. K. Sonoda, K. Nakano, H. Yamasita, N. Nakayama, and Y. Jinja, "Research on improvements in combustion of high water-content emulsified fuel, " in Proceedings of the 5th International Symposium on Marine Engineering (ISME '95), vol. 1, pp. 450–456, Yokohama, Japan, July 1995.

21. W. Remmels, A. Velji, R. -M. Schmidt, and M. Rauscher, "An experimental and theoretical study of exhaust gas recirculation in Diesel engines, " in Proceedings of 21st CIMAC Congress Interlaken (CIMAC '95), London, UK, 1995.

22. K. Yaguchi, T. Yoshida, K. Sato, T. Kobayashi, and A. Ishii, "High NOx reduction system mounted on 500 GT class vessel, " in Proceedings of the 5th International Symposium on Marine Engineering (ISME '95), vol. 1, Yokohama, Japan, July 1995.

Tribochemistry and EP Activity Assessment of Mo-S Complexes in Lithium-Base Greases

Tarunendr Singh

ABSTRACT

The blends of bis(1,5-diaryl-2,4-dithiomalonamido)dioxomolybdenum(VI) complexes in lithium-base grease are evaluated for their extreme pressure activity in a "four-ball test" using 12.7 mm diameter alloy steel ball specimen. The additive, bis(1,5-di-p-methoxyphenyl-2,4-dithiomalonamido)dioxomolybdenum(VI) and bis(1,5-di-p-chloro-phenyl-2,4-dithiomalonamido)dioxomolybdenum(VI) exhibited lower values of wear-scar diameter at higher load and higher values of weld load, flash temperature parameter, and pressure wear index as compared with lithium-base grease without additives. The greases fortified with the developed additives prevent rusting and corrosion of bearings while grease containing no additives did not pass these tests as per the standard tests. These greases have also better oxidation protection as compared to the grease that has no additive. The topography and tribochemistry of the wear-scar surface are carried out by means of scanning electron microscopy and Auger electron spectroscopy techniques, respectively.

Introduction

Lubricants are employed in bearings to decrease wear and destructive heating and to increase mechanical efficiency. Under extreme pressure, the boundary film of lubricant may fail leading to deterioration of surface properties causing adhesion, seizure, and excess wear. Using an appropriate extreme pressure, lubricant additive may minimize the friction and damage, which form a chemical film with the material of the surface. Certain lubricants containing EP additives find extensive applications in hypoid gears and in metal cutting and forming operations. The use of molybdenum disulphide as an excellent solid lubricant is well known, however, its insolubility in oil prevents its use in liquid lubricants. Oil-soluble organomolybdenum compounds, for example, molybdenum dialkyldithiophosphate (MoDTP) and molybdenum dialkyldithiocarbamate (MoDTC), have been used as excellent antifriction, antiwear, and extreme pressure additives in lubricating oils and greases. These additives that are known to increase the load-carrying capacity of lubricants and reduce fuel consumption and power loss by reducing friction lubricants containing oil-soluble organometallic compounds, sometimes referred as third-generation lubricants, have been known to show outstanding advantages over conventional lubricants in enhancing the component life, reducing operating temperatures and greatly extending lubrication intervals [1–8].

It is believed that the molybdenum and sulphur present in the additive form a low-friction surface film during operation of a machine under high loads. In view of these observations and our interest in searching for better extreme pressure additives, we are reporting, in this paper, few potential molybdenum-sulphur complexes as potential extreme pressure additives for lithium-base lubricating grease and their tribochemistry by AES and topography by SEM techniques.

Lithium-base greases are widely used in industrial machineries and automobiles [9]. These greases are formulated with various compounds with a view to optimizing the end use. In steel plant applications, lithium-12-hydroxy-stearate greases, formulated with extreme pressure additives, have performed well and they vary from sinter plant to coke oven. These greases have shown better performance, for example, high drop point, better antiwear and extreme pressure properties, better oxidation stability, pumpability as well as better wheel bearing performance as compared to conventional calcium base grease. The use of certain S-P and Pb-S systems in the lithium-base grease has been on record [10–13].

Experimental

Lithium Grease

Lithium-base grease (Grease A) was prepared in situ by reacting 12-hydroxystearic with lithium hydroxide monohydrate in presence of paraffinic mineral oil (see Table 1).

Table 1

	Test	Results	Methods
(1)	Nature of the base oil	Paraffinic mineral base oil	—
(2)	Base oil viscosity		ASTM D 445
	at 40°C	100.70	
	at 100°C	10.94	
(3)	Viscosity index	92	ASTM D 2270
(4)	Flash point (COC), °C	234	ASTM D 92
(5)	Pour point, °C	0	ASTM D 97
(6)	Aniline point, °C	104	ASTM D 611

Base Oil

Typical characteristics of the base oil used in the prepared lithium-base grease are given in Table 1.

Additives

The Mo-S complex additives were prepared [14, 15] by reacting an ethanolic solution of an appropriate 1,5-di-aryl-2,4-dithiomalonamide (2 mol) with an acidic solution of ammonium molybdate (1 mol) and digesting the reaction mixture for 15 minutes on a water bath.

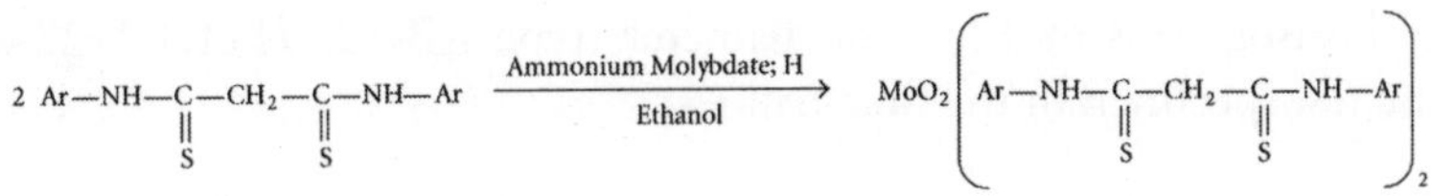

(1)

1,5-di-aryl-2,4-dithiomalonamide

Bis (1,5-diaryl-2,4-dithiomalonamido) dioxomolybdenum (VI)

Whereas, Ar=p-Methoxyphenyl- and p-Chlorophenyl- groups.

Preparation of Bis(1,5-diaryl-2,4-dithiomalonamido) dioxomolybdenum (VI)

The following two Mo-S complexes were prepared and used as extreme pressure additives in the lithium-base grease.

$$MoO_2\left(CH_3O{-}H_4C_6{-}NH{-}\underset{\underset{S}{\|}}{C}{-}CH_2{-}\underset{\underset{S}{\|}}{C}{-}NH{-}C_6H_4{-}OCH_3\right)_2.$$

$$MoO_2\left(Cl{-}H_4C_6{-}NH{-}\underset{\underset{S}{\|}}{C}{-}CH_2{-}\underset{\underset{S}{\|}}{C}{-}NH{-}C_6H_4{-}Cl\right)_2.$$

Grease Additive Admixture

The optimized dosages of the above additives were blended at a temperature of 80°C with the prepared Grease 1 (without additive) and with a view to achieve the following two main properties for the extreme pressure greases. (a)Timken OK load of 60 lbs minimum. (b)Four-ball weld load of 315 kgf minimum.

The following greases were used in this study:

Grease A: Lithium-base grease without additive.
Grease B: Grease A + bis(1,5-di-p-methoxyphenyl-2,4-dithiomalonamido)-dioxomolybdenum(VI).
Grease C: Grease A + bis(1,5-di-p-chlorophenyl-2,4-dithiomalonamido)-dioxomolybdenum(VI).

The rust and corrosion inhibitors were also incorporated in the above greases with a view to pass the following tests. (a) Emcor rust test, ASTM D 6138 [16]. (b) Corrosion prevention test, ASTM D 1743 [16].

Test Balls

SKF steel bearing balls of 12.7 mm diameter (type RB-12.7/111/E212) were used as the test specimen in the four-ball test.

Apparatus

(a) Extreme pressure lubricant test

(i)Four-ball machine. The tests were conducted on a four-ball machine by following standard procedure [16]. The duration of the test was 60 seconds. A series of tests were performed until the welding point was reached, and the determined parameters were initial seizure load (ISL), 2.5-second seizure delay load (SDL), just before weld load (JBWL), weld load (WL), wear-scar diameter (d) at ISL and JBWL, flash temperature parameter (FTP), and pressure wear index (PWI).

(ii)Timken machine. The measurement of load-carrying capacity of the prepared polyurea greases was carried out by Timken method by following ASTM D 2509 test method [16]. The duration of the test was ten minutes. The test results obtained with the Timken machine are recorded as Timken OK load in Table 3.

(b) Rust tests. These tests were performed by following ASTM D 6138 method in SKF Emcor test rig [16] and ASTM D 1743 test rig [16]. The prepared polyurea greases (Grease 2 and Grease 3) pass the rust tests as per ASTM D 6138 and ASTM D 1743.

(c) Oxidation stability test. It was carried out using oxidation bomb method, ASTM D 942, [16].

(d) Topography of the wear-scar surface was studied by scanning electron microscopy (Phillips XL-20) technique [17]. The wear scar obtained with Grease A, Grease B, and Grease C after "four-ball" test at just before weld loads were selected for the study. The stains of sludge or varnish found on the ball surface in the vicinity of the wear scar were removed with cotton and then cleaned with acetone in an ultrasonic bath before taking the micrographs.

(e) Tribochemistry was performed to study the film, formed on the wear-scar surface, obtained after "four-ball" test using AES technique [18]. The presence of sulphur and molybdenum in the film, formed during lubrication, was detected. The loads selected for the investigation were just before weld loads (for Grease B and Grease C). The tested ball surface was cleaned with acetone in an ultrasonic bath before the AES analyses.

Results and Discussion

The prepared greases were evaluated for the extreme pressure, rust and corrosion inhibition, and oxidation stability properties, and the results are reported in Tables 2, 3.

Table 2: Summary of the experimental results obtained with prepared Grease A, Grease B, and Grease C in the four-ball test.

Sl. No.	LUBRICANT	ISL* kgf	*d** at ISL mm	SDL* kgf	JBWL* kgf	*d** at JBWL* mm	WL* kgf	FTP* max	PWI* max
(1)	Grease A	80	0.70	100	140	1.90	160	132	9
(2)	Grease B	112	0.70	126	280	2.60	315	217	11
(3)	Grease C	112	0.50	126	280	2.50	315	270	11

*d** = wear-scar diameter, ISL* = initial seizure load, SDL* = 2.5-second seizure delay load, JBWL* = just before weld load, WL* = weld load, FTP* = flash temperature parameter, PWI* = pressure wear index.

Table 2 presents a record of the values of the initial seizure load (ISL), wear-scar diameter (d), 2.5-second seizure delay load (SDL), just before weld load (JBWL), weld load (WL), flash temperature parameter (FTP), and pressure wear index (PWI) for the prepared Grease A, Grease B, and Grease C in the four-ball test.

Table 3: Typical test results recorded for the prepared Grease A, Grease B, and Grease C.

Sl. No.	Test	Results GREASE A	Results GREASE B	Results GREASE C	Methods ASTM
(1)	NLGI grade	2	2	2	NLGI
(2)	Penetration, 60 X	274	272	272	D 217
	Penetration, 100,000 X	299	294	295	
(3)	Drop point, °C	198	198	199	D 2265
(4)	Oil separation, % by wt at 100°C, 30 h	1.4	1.5	1.8	D 6184
(5)	Oil separation on storage, % by wt	Nil	Nil	Nil	D 1742
(6)	Oxidation stab. at 99 ± 1°C, 100 h pressure drop, psi	15	2	1.5	D 942
(7)	Water washout at 80°C, % loss by wt	4.0	4.2	4.0	D 1264
(8)	Rust test	FAILS	1	1	D 1743
(9)	EMCOR rust test	FAILS	0,0	0,0	D 6138
(10)	Copper corrosion test at 100°C, 24 h	PASS	PASS	PASS	D 4048
(11)	Deleterious particles, number of scratches	NIL	NIL	NIL	D 1404
(12)	Wheel bearing test at 105°C for 6 h, leakage by wt, gms.	1.2	1.2	1.3	D 1263
(13)	Four-ball wear test, Wear-scar dia., mm, ~1200 rpm, 40 kg at75°C	0.85	0.45	0.42	D 2266
(14)	Four-ball weld load, kg	160	315	315	D 2596
(15)	Timken Ok Load, lbs	20	45	45	D 2509

A series of test were conducted with the prepared greases, a “four-ball machine”, and it was found that the wear-scar diameter values increase gradually up to ISL, owing to the physiosorbed/chemisorbed thin layers of the additive/lubricant on the rubbing surfaces. On increasing the load, there is a sharp increase in the wear-scar diameter values, indicating that the adsorbed lubricant/additive film has become partially desorbed. The transition from ISL and the consequent rise in the temperature from ISL to SDL result in the chemisorption of additive film on the contacting surfaces, and ultimately a chemical film of iron-sulphide/iron-oxide/molybdenum-disulphide is believed to be formed. This mixed film separates the contact surfaces and lower the friction coefficient and wear-scar diameter values, even at much higher loads.

Table 3 records the typical test results for the prepared greases (Grease A, Grease B, and Grease C). The additive fortified greases (Grease B and Grease C) show better oxidation stability as well as excellent rust and corrosion protection properties. The greases (Grease B and Grease C) show higher load carrying properties in four-ball test (weld loads of 315 kgf) and Timken test (Timken OK loads of 45 lbs) as compared to the Grease A (without additive).

Topography

The effectiveness of the prepared greases was examined by the technique of scanning electron microscopy [17]. The wear scar obtained with Grease A, Grease B, and Grease C after four-ball test at just before weld loads was selected for the study. The stains of sludge or oil found on the surface of the test specimens in the vicinity of the wear scar were removed with cotton. The tested specimens were then cleaned with acetone in an ultrasonic bath. The micrographs are shown in Figure 1. The micrographs of Grease B and Grease C exhibit smoother surfaces in comparison to the micrograph obtained with Grease A. The flow pattern at the leading edges indicates adhesive wear, which might be due to the junction growth and subsequent rupture. The black lines and spots are probably due to the formation of mixed chemical films of iron-sulphide/iron-oxide/molybdenum-disulphide, which provide effective lubrication at higher loads. Thus, the SEM study confirmed the wear scar obtained with Grease B and Grease C is smoother in nature as compared to the wear scar obtained with Grease A.

Tribochemistry

Auger electron spectroscopy [18] was used toexplore the chemical composition of the film formed during lubrication in the wear tracks. The "Auger spectra" were obtained by plotting the derivative of electron energy distribution against energy. The Auger analyses of the used balls were analyzed using scanning electron micro-probe. The Auger spectra obtained with line analyses for Grease B (see Figure 2) and Grease C (see Figure 3) (at just before weld load) show the presence of oxygen, carbon, sulphur, nitrogen, chlorine, and molybdenum. The quantitatively detected elements are absent in the lithium grease without additive (Grease A at just before weld load). It appears that the sulphur and molybdenum are responsible for higher weld load for Grease B and Grease C, this may probably be due to the formation of iron-sulphide as well as molybdenum-disulphide at high temperature, which diffuses into the first atomic layer of the metal and forms a new alloy, which then provides effective lubrication to the bearing ball surfaces.

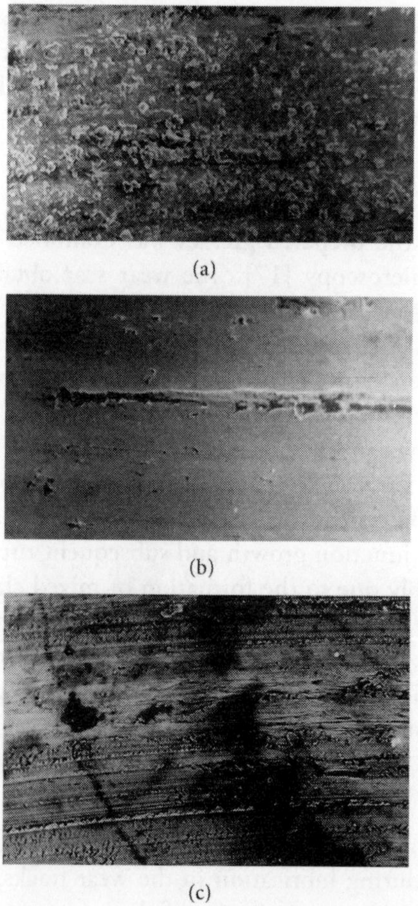

Figure 1: Scanning electron micrographs. (a) Grease A at 140 kgf, (b) Grease B at 280 kgf, (c) Grease C at 280 kgf.

Thus, it can be corroborated that these elements are derived from the additives which are used in the lithium-base grease and are responsible for the smooth efficiency of the prepared greases at higher loads.

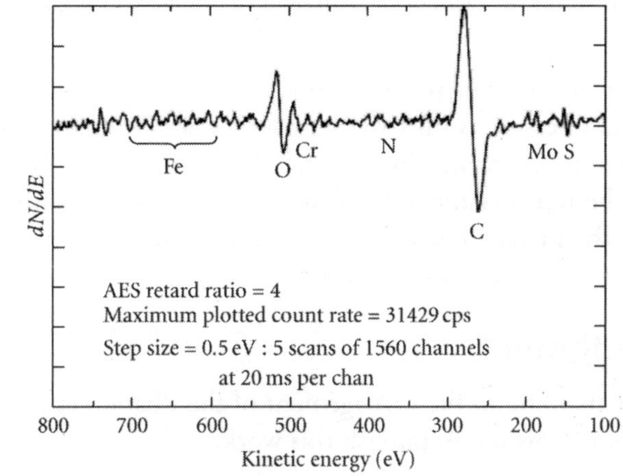

Figure 2: AES of Grease B at just before weld load; 280 kg.

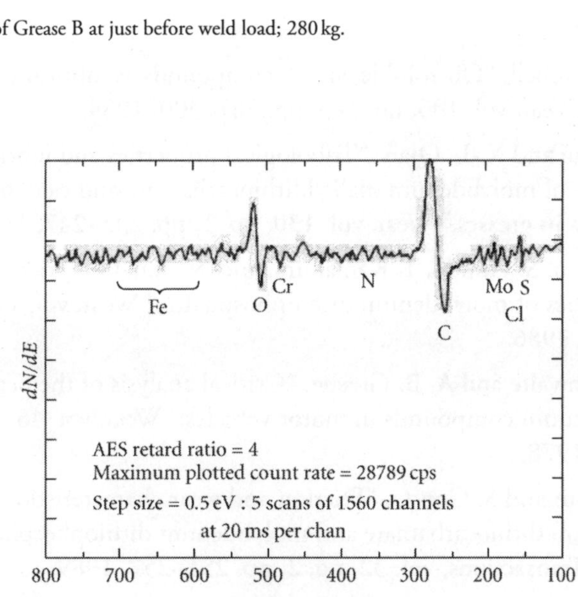

Figure 3: AES of Grease C at just before weld load; 280 kg.

Conclusions

The prepared greases blended with different additives were found to be more effective in reducing friction and wear at sliding surfaces, and increasing the load carrying capacities as compared to Grease A (grease without additives), Grease B, and Grease C showed lower values of wear-scar diameter and higher values of weld loads.

The scanning electron micrographs showed that the wear scar obtained with the Grease B and Grease C is smoother in nature than Grease A. The tribochemistry by AES analyses confirmed the presence of sulphur and molybdenum in the wear tracks of the test specimen obtained after the four-ball test with Grease B and Grease C. The prepared greases (Grease B and Grease C) also pass the rust and corrosion and oxidation stability tests.

Acknowledgment

The authors are thankful to the management of the Bharat Petroleum Corporation Limited for permission to publish this work.

References

1. P. C. H. Mitchell, "Oil-soluble MO-S compounds as lubricant additives," Wear, vol. 100, no. 1–3, pp. 281–300, 1984.
2. E.-C. Zheng and X.-L. Qian, "Tribological properties and lubrication mechanism of molybdenum dialkyldithiocarbamate and calcium compounds in greases," Wear, vol. 130, no. 1, pp. 233–247, 1989.
3. Y. Yamamoto, S. Gondo, T. Kamakura, and N. Tanaka, "Frictional characteristics of molybdenum dithiophosphates," Wear, vol. 112, no. 1, pp. 79–87, 1986.
4. E. R. Braithwaite and A. B. Greene, "Critical analysis of the performance of molybdenum compounds in motor vehicles," Wear, vol. 46, no. 2, pp. 405–431, 1978.
5. Y. Yamamoto and S. Gondo, "Friction and wear characteristics of molybdenum dithiocarbamate and molybdenum dithiophosphate," Tribology Transactions, vol. 32, no. 2, pp. 251–257, 1989.
6. K. Malsuo, Journal ofJSLE, vol. 31, no. 4, p. 260, 1986.
7. T. Singh and V. K. Verma, "New organomolybdenum compounds as potential E.P. lubricant additives," Indian Journal of Technology, vol. 28, no. 11, pp. 649–656, 1990.

8. A. B. Green and T. J. Ridson, "The effect of molybdenumcontaining, oil-soluble friction modifier on engine fuel economy and gear oil efficiency," SAE Technical paper 811187, 1981.
9. C. J. Bonor, "Manufacture and application of lubricating greases," National Lubricating Grease Institute, USA, 1994.
10. A. G. Izcus, "The effects of lead compounds upon the droppig points of lithium lubricating grease," NLGI Spokesman, vol. 44, no. 8, pp. 280–290, 1980.
11. J. R. Hasting, "Lithium and lithium complex greases for use in the steel industry," Lubrication Engineering, vol. 37, no. 2, pp. 91–94, 1981.
12. S. K. Sharma, P. Vasudevan, and U. S. Tewari, "High temperature lubricants-oils and greases," Tribology International, vol. 16, no. 4, pp. 213–219, 1983.
13. R.-H. Jiang, "Effects of the composition and fibrous texture of lithium soap grease on wear and friction," Tribology International, vol. 18, no. 2, pp. 121–124, 1985.
14. G. Barnikov, V. Kath, and D. Richter, "Isothiocyanate. II. N,N⌧-Aryl-substituierte Dithiomalonsäurediamide. I," Journal für Praktische Chemie, vol. 30, no. 1-2, pp. 63–66, 1965, Chemical Abstract, 64, 3411c, 1966.
15. T. Pal, P. Anjali, and P. K. Das, Journal of the Indian Chemical Society, vol. 65, p. 821, 1988.
16. Petroleum Products and Lubricants, Annual Book of ASTM Standards, Vol. 5.01-5.03.
17. M. Tomaru, S. Hironaka, and T. Sakurai, "Effects of oxygen on the load-carrying action of some additives," Wear, vol. 41, no. 1, pp. 117–140, 1977.
18. T. P. Debies and W. G. Johnston, "Surface chemistry of some antiwear additives as determined by electron spectroscopy," ASLE Transactions, vol. 23, no. 3, pp. 289–297, 1980.

GLOSSARY OF USEFUL TERMS

Absolute Pressure Transducer: A transducer which measures pressure in relation to zero pressure (a vacuum on one side of the diaphragm).

Absolute Pressure: Gage pressure plus atmospheric pressure.

Absolute Zero: Temperature at which thermal energy is at a minimum. Defined as 0 Kelvin, calculated to be -273.15°C or -459.67°F.

Acceleration: A change in the velocity of a body or particle with respect to time. The parameter that an accelerometer measures (dv/dt). Units expressed in "g".

Accelerometer: A device which converts the effects of mechanical motion into an electrical signal that is proportional to the acceleration value of the motion. A sensor. A transducer.

Accuracy: The closeness of an indication or reading of a measurement device to the actual value of the quantity being measured. Usually expressed as ± percent of full scale output or reading.

Acoustics: The degree of sound. The nature, cause, and phenomena of the vibrations of elastic bodies; which vibrations create compressional waves or wave fronts which are transmitted through various media, such as air, water, wood, steel, etc.

Activity (ai): A thermodynamic term for the apparent or active concentration of a free ion in solution. It is related to concentration by the activity coefficient.

Activity Coefficient (fi): A ratio of the activity of species i(ai) to its molality (C). It is a correction factor which makes the thermodynamic calculations correct. This factor is dependent on ionic strength, temperature, and other parameters.

Individual ionic activity coefficients, f+ for cation and f- for an anion, cannot be derived thermodynamically. They can be calculated only by using the Debye-Huckel law for low concentration solutions in which the interionic forces depend primarily on charge, radius, and distribution of the ions and on the dielectric constant of the medium rather than on the chemical properties of the ions.

Mean ionic activity coefficient (f±) or the activity of a salt, on the other hand, can be measured by a variety of techniques such as freezing point depression and vapor pressure as well as paired sensing electrodes. It is the geometric mean of the individual ionic activity coefficients

f± = (f+n+f-n-)1/n

Adapter: A mechanism or device for attaching non-mating parts.

ADC (Analog-to-Digital Converter):an electronic device which converts analog signals to an equivalent digital form, in either a binary code or a binary-coded-decimal code. When used for dynamic waveforms, the sampling rate must be high to prevent aliasing errors from occurring.

Additives: Chemicals added to fuel in very small quantities to improve and maintain fuel quality and/or to lower emissions.

Aftercooling / Intercooling: Cooling the engine intake air after the turbocharger and prior to introduction into the cylinder. Aftercooling increases engine power and lowers NOx emissions.

Aftertreatment Devices: Devices which remove pollutants from exhaust gases after the gas leaves combustion chamber (e.g., catalytic converters or diesel particulate filters). The term "exhaust gas aftertreatment" is considered derogatory by some in the emission control industry, but there is no consensus on the use of such alternatives as "post-combustion treatment" or "exhaust emission control".

Air Toxics: Toxic air pollutants, as classified by pertinent regulations. Examples of substances classified as air toxics by the US Clean Air Act include acetaldehyde, benzene, 1,3-butadiene, formaldehyde, and polycyclic organic matter (POM). California air toxics regulations also classify diesel exhaust particulates as a toxic air contaminant.

Aliasing: If the sample rate of a function (fs) is less than two times the highest frequency value of the function, the frequency is ambiguously presented. The frequencies above (fs/2) will be folded back into the lower frequencies producing erroneous data.

Alloy 11: A compensating alloy used in conjunction with pure copper as the negative leg to form extension wire for platinum-platinum rhodium thermocouples Types R and S.

Alloy 200/226: The combination of compensating alloys used with tungsten vs. tungsten 26% rhenium thermocouples as extension cable for applications under 200°C.

Alloy 203/225: The combination of compensating alloys used with tungsten 3% rhenium vs. tungsten 150 rhenium thermocouples as extension cable for applications under 200°C.

Alloy 405/426: The combination of compensating alloys used with tungsten 5% rhenium vs. tungsten 26% rhenium thermocouples as extension cable for applications under 870°C.

Alphanumeric: A character set that contains both letters and digits.

Alternative Fuel: Fuel other than petroleum diesel or gasoline.

Alumel: An aluminum nickel alloy used in the negative leg of a Type K thermocouple (Trade name of Hoskins Manufacturing Company).

Ambient Compensation: The design of an instrument such that changes in ambient temperature do not affect the readings of the instrument.

Ambient Conditions: The conditions around the transducer (pressure, temperature, etc.).

Ambient Pressure: Pressure of the air surrounding a transducer.

Ambient Temperature: The average or mean temperature of the surrounding air which comes in contact with the equipment and instruments under test.

Ammeter: An instrument used to measure current.

Ampere (amp): A unit used to define the rate of flow of electricity (current) in a circuit; units are one coulomb (6.28 x 1018 electronics) per second.

Amplifier: A device which draws power from a source other than the input signal and which produces as an output an enlarged reproduction of the essential features of its input.

Amplitude Span: The Y-axis range of a graphic display of data in either the time or frequency domain. Usually a log display (dB) but can also be linear.

Amplitude: A measurement of the distance from the highest to the lowest excursion of motion, as in the case of mechanical body in oscillation or the peak-to-peak swing of an electrical waveform.

Analog Output: A voltage or current signal that is a continuous function of the measured parameter.

Analog-to-Digital Converter (A/D or ADC): A device or circuit that outputs a binary number corresponding to an analog signal level at the input.

Anemometer: An instrument for measuring and/or indicating the velocity of air flow.

Angstrom: Ten to the minus tenth meters (10-10) or one millimicron, a unit used to define the wave length of light. Designated by the symbol ‰.

Angular Frequency: The motion of a body or a point moving circularly, referred to as the circular frequency O which is the frequency in cycles per second (cps) multiplied by the term (2) and expressed in radians per second (2pf).

Anion: A negatively charged ion (Cl-, NO3-, S2- etc.)

ANSI: American National Standards Institute.

Anti-reset Windup: This is a feature in a three-mode PID controller which prevents the integral (auto reset) circuit from functioning when the temperature is outside the proportional band.

Application Program: A computer program that accomplishes specific tasks, such as word processing.

Articulated Pistons: Two-piece pistons incorporating an entirely separate piston crown or dome with a separate skirt, and linking the two together with the piston pin. Many 1994 and later engines incorporate steel crown/aluminum skirt articulated pistons.

ASME: American Society of Mechanical Engineers.

Assembler: A program that translates assembly language instructions into machine language instructions.

Assembly Language: A machine oriented language in which mnemonics are used to represent each machine language instruction. Each CPU has its own specific assembly language.

ASTM: American Society for Testing and Materials.

Asymmetry Potential: The potential developed across the glass membrane with identical solutions on both sides. Also a term used when comparing glass electrode potential in pH 7 buffer.

Asynchronous: A communication method where data is sent when it is ready without being referenced to a timing clock, rather than waiting until the receiver signals that it is ready to receive.

ATC: Automatic temperature compensation.

Auto-Zero: An automatic internal correction for offsets and/or drift at zero voltage input.

Automatic Reset: 1. A feature on a limit controller that automatically resets the controller when the controlled temperature returns to within the limit bandwidth set. 2. The integral function on a PID controller which adjusts the proportional bandwidth with respect to the set point to compensate for droop in the circuit, i.e., adjusts the controlled temperature to a set point after the system stabilizes.

AWG: American Wire Gage.

Axis of Rotation (Spin Axis): The axis of rotation (spin axis) is that straight line about which a body rotates.

Background Noise: The total noise floor from all sources of interference in a measurement system, independent of the presence of a data signal.

Backup: A system, device, file or facility that can be used as an alternative in case of a malfunction or loss of data.

Bandwidth: A symmetrical region around the set point in which proportional control occurs.

Bearing: A part which supports a journal and in which a journal revolves.

Beat Frequency: Beat frequencies are periodic vibrations that result from the addition and subtraction of two or more sinusoids. For example, in the case of two turbine aircraft engines that are rotating at nearly the same frequency but not precisely at the same frequency; Four frequencies are generated:(f1) the rotational frequency of turbine one, (f2) the rotational frequency of turbine two, (f1 + f2) the sum of turbine rotational frequencies one and two, and (f1 - f2) which is the difference or "beat" frequency of turbines one and two. The difference of the two frequencies is the lower frequency and is the one that is "felt" as a beat or "wow" in this case.

Beryllia: BeO (Beryllium Oxide) A high-temperature mineral insulation material; toxic when in powder form.

Best Fit Straight Line (BFSL): A line midway between two parallel straight lines enclosing all output vs. pressure values.

Beta Ratio: The ratio of the diameter of a pipeline constriction to the unconstricted pipe diameter.

Bi-Fueled Vehicle: A vehicle with two separated fuel systems designed to run on either conventional fuel or an alternative fuel using only one fuel at a time.

Biodiesel: The mono alkyl esters of long chain fatty acids derived from renewable lipid feedstocks, such as vegetable oils and animal fats, for use in compression ignition (diesel) engines. Manufactured by transestrification of the organic feedstock by methanol.

Boiling Point: The temperature at which a substance in the liquid phase transforms to the gaseous phase; commonly refers to the boiling point of water which is 100°C (212°F) at sea level.

Brake Mean Effective Pressure (BMEP): The work accomplished during one engine cycle divided by the engine swept volume. It is essentially the engine torque normalized by the engine displacement. The word "brake" denotes the actual torque/power available at the engine flywheel as measured on a dynamometer. Thus, BMEP is a measure of the useful power output of the engine.

Brake Specific Fuel Consumption (BSFC): BSFC is the ratio of the engine fuel consumption to the engine power output (as measured at the flywheel). BSFC has units of grams of fuel per kilowatt-hour (g/kWh) or pounds mass of fuel per brake horsepower-hour (lb/bhp-hr). BSFC is a measure of engine efficiency.

Bridge Resistance: See Input impedance and Output impedance.

BTU: British thermal units. The quantity of thermal energy required to raise one pound of water at its maximum density, 1 degree F. One BTU is equivalent to .293 watt hours, or 252 calories. One kilowatt hour is equivalent to 3412 BTU.

Buffer Capacity (B): A measure of the ability of the solution to resist pH change when a strong acid or base is added.

Buffer: 1. A storage area for data that is used to compensate for a speed difference, when transferring data from one device to another. Usually refers to an area reserved for I/O operations, into which data is read, or from which data is written.2. Any substance or combination of substances which, when dissolved in water, produces a solution which resists a change in its hydrogen ion concentration on the addition of an acid or alkali.

Bulb (Liquid-in-Glass Thermometer): The area at the tip of a liquid-in-glass thermometer containing the liquid reservoir.

Burst Pressure: The maximum pressure applied to a transducer sensing element or case without causing leakage.

Burst Proportioning: A fast-cycling output form on a time proportioning controller (typically adjustable from 2 to 4 seconds) used in conjunction with a solid state relay to prolong the life of heaters by minimizing thermal stress.

Bus: Parallel lines used to transfer signals between devices or components. Computers are often described by their bus structure (i.e., S-100, IBM PC).

Calender-van Dusen Equation: An equation that defines the resistance-temperature value of any pure metal that takes the form of RT = RO(1 + AT + BT2) for values between the ice point (0°C) and the freezing point of antimony (630.7°C) and the form RT = RO[1 + AT + BT2 + C(T-100)T2] between the oxygen point (-183.0°C) and the ice point (0°C).

Calibration: The process of adjusting an instrument or compiling a deviation chart so that its reading can be correlated to the actual value being measured.

Calorie: The quantity of thermal energy required to raise one gram of water 1°C at 15°C.

Carbon Dioxide (CO2): A colorless, odorless, non-toxic gas. It is one of main products of fossil-fuel combustion. Carbon dioxide is a greenhouse gas that contributes to the potential for global warming.

Carbon Monoxide (CO): A colorless, odorless and toxic gas. It blocks the lungs' ability to obtain oxygen. CO is produced by incomplete combustion of fossil fuels and is a major part of air pollution. Compression ignition (diesel) engines generate significantly lower CO emissions than spark ignited engines.

Carcinogens: Substances known to cause cancer.

Catalyst: A substance which influences the rate of a chemical reaction but is not one of the original reactants or final products, i.e. it is not consumed or altered in the reaction. Catalysts are used in many processes in the chemical and petroleum industries. Emission control catalysts are used to promote reactions that change exhaust pollutants from internal combustion engines into harmless substances.

Cation: A positively charged ion (Na+, H+).

Cavitation: The boiling of a liquid caused by a decrease in pressure rather than an increase in temperature.

Celsius (centrigrade): A temperature scale defined by 0°C at the ice point and 100°C at boiling point of water at sea level.

Center of Gravity (Mass Center): The center of gravity of a body is that point in the body through which passes the resultant of weights of its component particles for all orientations of the body with respect to a uniform gravitational field.

Centripetal Force: A force exerted on an object moving in a circular path which is exerted inward toward the center of rotation.

Ceramic Insulation: High-temperature compositions of metal oxides used to insulate a pair of thermocouple wires The most common are Alumina (Al2O3), Beryllia (BeO), and Magnesia (MgO). Their application depends upon temperature and type of thermocouple. High-purity alumina is required for platinum alloy thermocouples. Ceramic insulators are available as single and multihole tubes or as beads.

Ceramic: Polycrystalline ferroelectric materials which are used as the sensing units in piezoelectric accelerometers. There are many different grades, all of which can be made in various configurations to satisfy different design requirements.

Cetane Index: A calculated value, derived from fuel density and volatility, giving a reasonably close approximation to cetane number.

Cetane Number: A measure of ignition quality of diesel fuel. The higher the cetane number the easier the fuel ignites when injected into an engine. Cetane number is determined by an engine test using two reference fuel blends of known cetane numbers. The reference fuels are prepared by blending normal cetane (n-hexadecane), having a value of 100, with heptamethyl nonane, having a value of 15.

CFM: The volumetric flow rate of a liquid or gas in cubic feet per minute.

CFR (Cooperative Fuel Research) Engine: A single cylinder, overhead valve, variable compression ratio engine used for measuring octane or cetane quality.

Charge Sensitivity: For accelerometers that are rated in terms of charge sensitivity, the output voltage (V)is proportional to the charge (Q) divided by the shunt capacitance (C). This type of accelerometer is characterized by a high output impedance. The sensitivity is given in terms of charge; picocoulombs per unit of acceleration (g).

Chatter: The rapid cycling on and off of a relay in a control process due to insufficient bandwidth in the controller.

Clear: To restore a device to a prescribed initial state, usually the zero state.

Clock: The device that generates periodic signals for synchronization.

Closeness of Control: Total temperature variation from a desired set point of system. Expressed as "closeness of control" is ±2°C or a system bandwidth with 4°C, also referred to as amplitude of deviation.

Cloud Point (CP): A measure of the ability of a diesel fuel to operate under cold weather conditions. Defined as the temperature at which wax first becomes visible when diesel fuel is cooled under standardized test conditions (ASTM D2500).

Cold Filter Plugging Point (CFPP): A measure of the ability of a diesel fuel to operate under cold weather conditions. Defined as the lowest temperature at which diesel fuel will pass through a fine wire mesh screen of the test apparatus.

Color Code: The ANSI established color code for thermocouple wires in the negative lead is always red. Color Code for base metal thermocouples is yellow for Type K, black for Type J, purple for Type E and blue for Type T.

Common Mode: The output form or type of control action used by a temperature controller to control temperature, i.e. on/off, time proportioning, PID.

Common Rail Injection: A diesel fuel injection system employing a common pressure accumulator, called the rail, which is mounted along the engine block. The rail is fed by a high pressure fuel pump. The injectors, which are fed from the common rail, are activated by solenoid valves. The solenoid valves and the fuel pump are electronically controlled. In the common rail injection system the injection pressure is independent from engine speed and load. Therefore, the injection parameters can be freely controlled. Usually a pilot injection is introduced, which allows for reductions in engine noise and NOx emissions.

Compensated Connector: A connector made of thermocouple alloys used to connect thermocouple probes and wires.

Compensating Alloys: Alloys used to connect thermocouples to instrumentation. These alloys are selected to have similar thermal electric properties as the thermocouple alloys (however, only over a very limited temperature range).

Compensation: An addition of specific materials or devices to counteract a known error.

Compressed Natural Gas (CNG): Natural gas compressed to a volume and density that is practical as a portable fuel supply.

Compression Ignition (CI): The form of ignition that initiates combustion in a diesel engine. The rapid compression of air within the cylinders generates the heat required to ignite the fuel as it is injected.

Conduction: The conveying of electrical energy or heat through or by means of a conductor.

Confidence Level: The range (with a specified value of uncertainty, usually expressed in percent) within which the true value of a measured quantity exists.

Conformity Error: For thermocouples and RTDs, the difference between the actual reading and the temperature shown in published tables for a specific voltage input.

Connection Head: An enclosure attached to the end of a thermocouple which can be cast iron, aluminum or plastic within which the electrical connections are made.

Constantan: A copper-nickel alloy used as the negative lead in Type E, Type J, and Type T thermocouples.

Continuous Spectrum: A frequency spectrum that is characterized by non-periodic data. The spectrum is continuous in the frequency domain and is characterized by an infinite number of frequency components.

Control Character: A character whose occurrence in a particular context starts, modifies or stops an operation that effects the recording, processing, transmission or interpretation of data.

Control Mode: The output form or type of control action used by a temperature controller to control temperature, i.e., on/off, time proportioning, PID.

Control Point: The temperature at which a system is to be maintained.

Convection: 1. The circulatory motion that occurs in a fluid at a non-uniform temperature owing to the variation of its density and the action of gravity. 2. The transfer of heat by this automatic circulation of fluid.

Converted Vehicle: A vehicle, originally designed to operated on gasoline or diesel, that has been modified or altered to operate on an alternative fuel.

Cordierite: A ceramic material of the formula 2MgO-2Al2O3-5SiO2 which is used for automotive flow-through catalyst substrates and ceramic wall-flow diesel filters.

Coriolis Force: A result of centripetal force on a mass moving with a velocity radially outward in a rotating plane.

Correction (Balancing) Plane: A plane perpendicular to the shaft axis of a rotor in which correction for unbalance is made.

Coulomb Sensitivity: Charge/unit acceleration, expressed in Pc/g (charge sensitivity).

Coulomb: A measurement of the quantity of electrical charge, usually expressed as pico coulomb (10-12 coulombs).

Counter Weight: A weight added to a body so as to reduce a calculated unbalance at a desired place.

Counts: The number of time intervals counted by the dual-slope A/D converter and displayed as the reading of the panel meter, before addition of the decimal point.

CPS: Cycles per second; the rate or number of periodic events in one second, expressed in Hertz (Hz).

Critical Damping: Critical damping is the smallest amount of damping at which a given system is able to respond to a step function without overshoot.

Critical Speed: The rotational speed of the rotor or rotating element at which resonance occurs in the system. The shaft speed at which at least one of the "critical" or natural frequencies of a shaft is excited.

Cryogenics: Measurement of temperature at extremely low values, i.e., below -200°C.

CSA: Canadian Standards Administration.

Cure Point: The temperature at which a normally magnetic material goes through a magnetic transformation and becomes non-magnetic.

Current Proportioning: An output form of a temperature controller which provides a current proportional to the amount of control required. Normally is a 4 to 20 milliamp current proportioning band.

Current: The rate of flow of electricity. The unit of the ampere (A) defined as 1 ampere = 1 coulomb per second.

Curve Fitting: Curve fitting is the process of computing the coefficients of a function to approximate the values of a given data set within that function. The approximation is called a "fit". A mathematical function, such as a least squares regression, is used to judge the accuracy of the fit.

Cycle Time: The time usually expressed in seconds for a controller to complete one on/off cycle.

Damping: The reduction of vibratory movement through dissipation of energy. Types include viscous, coulomb, and solid.

dB (Decibel): 20 times the log to the base 10 of the ratio of two voltages. Every 20 dBs correspond to a voltage ratio of 10, every 10 dBs to a voltage ratio of 3.162. For instance, a CMR of 120 dB provides voltage noise rejection of 1,000,000/1. An NMR of 70 dB provides voltage noise rejection of 3,162/1.

DC: Direct current; an electric current flowing in one direction only and substantially constant in value.

Dead Band: 1. For chart records the minimum change of input signal required to cause a deflection in the pen position. 2. For temperature controllers the temperature band where heat is turned off upon rising temperature and turned on upon falling temperature expressed in degrees. The area where no heating (or cooling) takes place.

Dead Volume: The volume of the pressure port of a transducer at room temperature and ambient barometric pressure.

Decimal: Refers to a base ten number system using the characters 0 through 9 to represent values.

Dedicated Vehicle: A vehicle designed to operate solely on one alternative fuel, as opposed to a converted vehicle which was later altered to use an alternative fuel.

Default: The value(s) or option(s) that are assumed during operation when not specified.

Degree: An incremental value in the temperature scale, i.e., there are 100 degrees between the ice point and the boiling point of water in the Celsius scale and 180°F between the same two points in the Fahrenheit scale.

Density: Mass per unit of volume of a substance, i.e., grams/cu.cm. or pounds/cu.ft.

Derivative: The derivative function senses the rate of rise or fall of the system temperature and automatically adjusts the cycle time of the controller to minimize overshoot or undershoot.

Deviation: The difference between the value of the controlled variable and the value at which it is being controlled.

Diaphragm: The sensing element consisting of a membrane which is deformed by the pressure differential applied across it.

Dielectric Constant: Related to the force of attraction between two opposite charges separated by a distance in a uniform medium.

Diesel Oxidation Catalyst (DOC): Catalyst promoting oxidation processes in diesel exhaust. Usually designed to reduce emissions of the organic fraction of diesel particulates, gas-phase hydrocarbons, and carbon monoxide.

Diesel Particulate Filter (DPF): A device which physically captures diesel particulates preventing their discharge from the tailpipe. Collected particulates need to be removed from the filter, usually by continuous or periodic oxidation in a process called "regeneration".

Diesel Particulate Matter (DPM): Sub-micron size particles found in diesel exhaust. Most emission regulations specify DPM measurement methods in which particulates are sampled on filters from cooled exhaust gas. The cooling causes condensation of vapors in the gas sampling train. Thus, the DPM is composed of both solid and liquid particles and is generally classified into three fractions: (1) inorganic carbon (soot), (2) organic fraction (often referred to as SOF or VOF), and (3) sulfate fraction (hydrated sulfuric acid).

Differential Pressure: The difference in static pressure between two identical pressure taps at the same elevation located in two different locations in a primary device.

Differential: For an on/off controller, it refers to the temperature difference between the temperature at which the controller turns heat off and the temperature at which the heat is turned back on. It is expressed in degrees.\

Digit: A measure of the display span of a panel meter. By convention, a full digit can assume any value from 0 through 9, a 1/2-digit will display a 1 and overload at 2, a 3/4-digit will display digits up to 3 and overload at 4, etc. For example, a meter with a display span of ±3999 counts is said to be a 3-3/4 digit meter.

Dimethyl Ether (DME): The simplest ether CH3-O-CH3. Can be manufactured from natural gas or from a renewable organic feedstock. DME is a prospective alternative diesel fuel.

DIN (Deutsche Industrial Norm): A set of German standards recognized throughout the world. The 1/8 DIN standard for panel meters specifies an outer bezel dimension of 96 x 48 mm and a panel cutout of 92 x 45 mm.

DIN 43760: The standard that defines the characteristics of a 100 ohm platinum RTD having a resistance vs. temperature curve specified by a = 0.00385 ohms per degree.

Direct Injection (DI): In diesel engines with direct injection the combustion chamber is not divided and fuel is injected directly to the cylinder.

Displacement: The measured distance traveled by a point from its position at rest. Peak to peak displacement is the total measured movement of a vibrating point between its positive and negative extremes. Measurement units expressed as inches or millinches.

Dissipation Constant: The ratio for a thermistor which relates a change in internal power dissipation to a resultant change of body temperature.

Dissociation Constant (K): A value which quantitatively expresses the extent to which a substance dissociates in solution. The smaller the value of K, the less dissociation of the species in solution. This value varies with temperature, ionic strength, and the nature of the solvent.

Drift: A change of a reading or a set point value over long periods due to several factors including change in ambient temperature, time, and line voltage.

Droop: A common occurrence in time-proportional controllers. It refers to the difference in temperature between the set point and where the system temperature actually stabilizes due to the time-proportioning action of the controller.

Dual Element Sensor: A sensor assembly with two independent sensing elements.

Dual-Fuel Vehicle: A vehicle designed to operate on a combination of alternative fuel, such as compressed natural gas (CNG) or liquefied petroleum gas (LPG), and conventional fuel, such as diesel or gasoline. These vehicles have two separate fuel systems, which inject both fuels simultaneously into the engine combustion chamber.

Dual-slope A/D Converter: An analog-to-digital converter which integrates the signal for a specific time, then counts time intervals for a reference voltage to bring the integrated signal back to zero. Such converters provide high resolution at low cost, excellent normal-mode noise rejection, and minimal dependence on circuit elements.

Duty Cycle: The total time to one on/off cycle. Usually refers to the on/off cycle time of a temperature controller.

Dynamic (Two-Plane) Balancing Machine: A dynamic balancing machine is a centrifugal balancing machine that furnishes information for performing two-plane balancing.

Dynamic Calibration: Calibration in which the input varies over a specific length of time and the output is recorded vs. time.

Dynamic Pressure: The difference in pressure levels from static pressure to stagnation pressure caused by an increase in velocity. Dynamic pressure increases by the square of the velocity.

Dynamic Unbalance: Dynamic unbalance is that condition in which the central principal axis is not coincident with the shaft axis.

Electromotive Force (emf): The potential difference between the two electrodes in a cell. The cell emf is the cell voltage measured when no current is flowing through the cell. It can be measured by means of a pH meter with high input impedance.

Electronic Control Module (ECM): A microprocessor that determines the beginning and end of each injection cycle on every cylinder. The ECM determines both fuel metering and injection timing in response to such parameters as engine crankshaft position and rpm, engine coolant and intake air temperature, and absolute intake air boost pressure.

Elemental Carbon (EC): Inorganic carbon, as opposed to carbon in organic compounds, sometimes used as a surrogate measure for diesel particulate matter, especially in occupational health environments. Ele-mental carbon usually accounts for 40-60% of the total DPM mass.

Endothermic: Absorbs heat. A process is said to be endothermic when it absorbs heat.

Enthalpy: The sum of the internal energy of a body and the product of its volume multiplied by the pressure.

Environmental Conditions: All conditions in which a transducer may be exposed during shipping, storage, handling, and operation.

Eutectic Temperature: The lowest possible melting point of a mixture of alloys.

Evaporative Emissions: Hydrocarbon vapors that escape from a fuel storage tank or a vehicle fuel tank or vehicle fuel system.

Exothermic: Gives off heat. A process is said to be exothermic when it releases heat.

Expansion Factor: Correction factor for the change in density between two pressure measurement areas in a constricted flow.

Explosion-proof Enclosure: An enclosure that can withstand an explosion of gases within it and prevent the explosion of gases surrounding it due to sparks, flashes or the explosion of the container itself, and maintain an external temperature which will not ignite the surrounding gases.

Exposed Junction: A form of construction of a thermocouple probe where the hot or measuring junction protrudes beyond the sheath material so as to be fully exposed to the medium being measured. This form of construction usually gives the fastest response time.

Fahrenheit: A temperature scale defined by 32° at the ice point and 212° at the boiling point of water at sea level.

Ferrule: A compressible tubular fitting that is compressed onto a probe inside a compression fitting to form a gas-tight seal.

Field Balancing Equipment: An assembly of measuring instruments for performing balancing operations on assembled machinery which is not mounted in a balancing machine.

Field of View: A volume in space defined by an angular cone extending from the focal plane of an instrument.

Flash Point: The temperature at which a combustible liquid gives off just enough vapor to produce a vapor/air mixture that will ignite when a flame is applied. The flash point is measured in a standardized apparatus using standard test methods, such as ASTM D93 or ISO 2719.

Flexible-Fueled Vehicle: A vehicle with the ability to operate on alternative fuels, 100 percent petroleum-based fuels, or a mixture of alternative fuel and petroleum-based fuels.

Flow Rate: Actual speed or velocity of fluid movement.

Flow: Travel of liquids or gases in response to a force (i.e. pressure or gravity).

Flowmeter: A device used for measuring the flow or quantity of a moving fluid.

FM Approved: An instrument that meets a specific set of specifications established by Factory Mutual Research Corporation.

FM: Factory Mutual Research Corporation. An organization which sets industrial safety standards.

Forced Vibration: Vibration of a system caused by an imposed force. Steady-state vibration is an unchanging condition of periodic or random motion.

Fossil Fuel: Any naturally occurring organic fuel formed in the Earth's crust, such as petroleum, coal and natural gas.

FPM: Flow velocity in feet per minute.

FPS: Flow velocity in feet per second.

Freezing Point: The temperature at which the substance goes from the liquid phase to the solid phase.

Frequency Modulated Output: A transducer output which is obtained in the form of a deviation from a center frequency, where the deviation is proportional to the applied stimulus.

Frequency of Vibration: The number of cycles occurring in a given unit of time. RPM - revolutions per minute. CPM- cycles per minute.

Frequency Output: An output in the form of frequency which varies as a function of the applied input.

Frequency, Natural: The frequency of free (not forced) oscillations of the sensing element of a fully assembled transducer.

Frequency: The number of cycles over a specified time period over which an event occurs. The reciprocal is called the period.

Fuel cycle: The processes involved in extracting a fuel in its native form, converting it to a useful product, transporting it to market, and consuming it at its final destination.

Full Bridge: A Wheatstone bridge configuration utilizing four active elements or strain gages.

Full Scale Output: The algebraic difference between the minimum output and maximum output.

Functions: Three mode PID controller. A timeproportioning controller with integral and derivative functions. The integral function automatically adjusts the system temperature to the set point temperature to eliminate droop due to the time proportioning function.

g: The force of acceleration due to gravity equal to 32.1739 ft/sec2 or 386 in./sec2.

Gage Factor: A measure of the ratio of the relative change of resistance to the relative change in length of a piezoresistive strain gage.

Gage Length: The distance between two points where the measurement of strain occurs.

Gage Pressure Transducer: A transducer which measures pressure in relation to the ambient pressure.

Gage pressure: Absolute pressure minus local atmospheric pressure.

Gas Turbine: A turbine powered by a gas, such as the hot gas produced in combustion processes.

Gas Turbine Engine: A combustion engine utilizing a turbine, such as a jet engine.

Genset: Generator set, a generating system comprising a combustion engine driving an electrical generator.

Geometric Surface Area (GSA): In monolith catalyst substrates, the total channel surface area per unit of substrate volume.

GPH: Volumetric flow rate in gallons per hour.

GPM: Volumetric flow rate in gallons per minute.

Gross Vehicle Weight (GVW): The maximum loaded weight (including the vehicle itself, passengers, and cargo) for which a vehicle is designed, as specified by the manufacturer. Often used as a criterion of vehicle size for the purpose of legislation; the exact definition may vary depending on the jurisdiction.

Half Bridge: Two active elements or strain gages.

Head Loss: The loss of pressure in a flow system measured using a length parameter (i.e., inches of water, inches of mercury).

Head Pressure: Pressure in terms of the height of fluid, P = yrg, where r = fluid density and y = the fluid column heights. Expression of a pressure in terms of the height of fluid, r = yrg, where r is fluid density and y = the fluid column height. g = the acceleration of gravity.

Heat Sink: 1. Thermodynamic. A body which can absorb thermal energy. 2. Practical. A finned piece of metal used to dissipate the heat of solid state components mounted on it.

Heat Transfer: The process of thermal energy flowing from a body of high energy to a body of low energy. Means of transfer are conduction; the two bodies contact. Convection; a form of conduction where the two bodies in contact are of different phases, i.e. solid and gas. Radiation all bodies emit infrared radiation.

Heat Treating: A process for treating metals where heating to a specific temperature and cooling at a specific rate changes the properties of the metal.

Heat: Thermal energy. Heat is expressed in units of calories or BTU's.

Hertz (Hz): Units in which frequency is expressed. Synonymous with cycles per second.

Hexadecimal: Refers to a base sixteen number system using the characters 0 through 9 and A through F to represent the values. Machine language programs are often written in hexadecimal notation.

Hooke's Law: Defines the basis for the measurement of mechanical stresses via the strain measurement. The gradient of Hooke's line is defined by the ratio of which is equivalent to the Modulus of Elasticity E (Young's Modulus).

Hybrid Electric Vehicle (HEV): Hybrid-electric vehicle. Various types of electric vehicles that use another power source to propel the vehicle or generate power for an electric drive train, or a combination of the two types.

Hydraulic/Electronic Unit Injector (HEUI): A type of unit injector actuated by engine oil pressure rather than the camshaft. A very high oil pressure (up to 3,000 psi) is created by a separate oil pump. This high pressure is routed to every injector through a gallery. The engine's Electronic Control Module varies the pressure in response to engine speed and other parameters.

Hydromechanical Injection: An injection system in which mechanical parts work through hydraulic pressure to meter and time the injection of fuel. No electronics are incorporated into hydromechanical injection systems.

Ignition Delay: The length of time or number of degrees of crankshaft rotation between the beginning of injection and ignition of the fuel.

In-Direct Injection (IDI): In diesel engines with in-direct injection the fuel is injected to an auxiliary pre-chamber. Combustion starts in the prechamber and propagates to the cylinder.

Infrared: An area in the electromagnetic spectrum extending beyond red light from 760 nanometers to 1000 microns (106 nm). It is the form of radiation used for making non-contact temperature measurements.

Initial Unbalance: Initial unbalance is that unbalance of any kind that exists in the rotor before balancing.

Injection Period: The time, measured in degrees of crankshaft rotation, between the beginning and end of injection. On engines with hydromechanical injection systems, it is controlled by the opening and closing of ports in the injector body or by the action of a plunger forcing fuel out of a cup. On electronic injection systems, it is determined directly or indirectly by the action of a solenoid valve.

In-Line Injection Pump: An injection pump with a separate cylinder and plunger for each engine cylinder. Each plunger is rotated by a rack to determine metering via ports in the body of the pump and helical cuts on the pump plungers. The plungers are driven off a camshaft, which usually incorporates a centrifugal or electronically controlled timing advance mechanism.

Insulated Junction: See Ungrounded Junction

Insulation Resistance: The resistance measured between two insulated points on a transducer when a specific dc voltage is applied at room temperature.

Interchangeability Error: A measurement error that can occur if two or more probes are used to make the same measurement. It is caused by a slight variation in characteristics of different probes.

Interface: The means by which two systems or devices are connected and interact with each other.

Interrupt: To stop a process in such a way that it can be resumed.

Intrinsically Safe: An instrument which will not produce any spark or thermal effects under normal or abnormal conditions that will ignite a specified gas mixture.

Intumescent Mat: Ceramic fiber mat which irreversibly expands after exposed to high temperature. Usually contains vermiculite. Intumescent mats are used in the canning of catalytic converters and diesel filters to hold the ceramic substrate inside the steel canister.

In-Use Deterioration (IUD): The effect of time and use on vehicle performance and emissions.

IPTS-48: International Practical Temperature Scale of 1948. Fixed points in thermometry as specified by the Ninth General Conference of Weights and Measures which was held in 1948.

IPTS-68: International Practical Temperature Scale of 1968. Fixed points in thermometry set by the 1968 General Conference of Weights and Measures.

ISA: Instrument Society of America.

Isolation: The reduction of the capacity of a system to respond to an external force by use of resilient isolating materials.

Isopotential Point: A potential which is not affected by temperature changes. It is the pH value at which dE/dt for a given electrode pair is zero. Normally, for a glass electrode and SCE reference, this potential is obtained approximately when immersed in pH 7 buffer.

Isothermal: A process or area that is a constant temperature.

Joule: The basic unit of thermal energy.

Journal: A journal is that part of a rotor that is in contact with or supported by a bearing in which it revolves.

Junction: The point in a thermocouple where the two dissimilar metals are joined.

Kelvin: Symbol K. The unit of absolute or thermodynamic temperature scale based upon the Celsius scale with 100 units between the ice point and boiling point of water. 0°C = 273.15K (there is no degree (°) symbol used with the Kelvin scale).

Kilowatt (kw): Equivalent to 1000 watts.

Kilowatt Hour (kwh): 1000 watthours. Kilovolt amperes (kva)

1000 volt amps.

Kinetic Energy: Energy associated with mass in motion, i.e., 1/2 rV2 where r is the density of the moving mass and V is its velocity.

KVA: Kilovolt amperes (1000-volt amps).

Lag: 1. A time delay between the output of a signal and the response of the instrument to which the signal is sent. 2. A time relationship between two waveforms where a fixed reference point on one wave occurs after the same point of the reference wave.

Laminar Flow: Streamlined flow of a fluid where viscous forces are more significant than inertial forces, generally below a Reynolds number of 2000.

Latent Heat: Expressed in BTU per pound. The amount of heat needed (absorbed) to convert a pound of boiling water to a pound of steam.

Leakage Rate: The maximum rate at which a fluid is permitted or determined to leak through a seal. The type of fluid, the differential Limits of Error

Lean NOx Catalyst (LNC): Catalyst designed to reduce nitrogen oxides from diesel or spark-ignited engine exhaust gases under net oxidizing conditions, i.e., in the presence of excessive amount of oxygen.

Least-squares Line: The straight line for which the sum of the squares of the residuals (deviations) is minimized.

Life Cycle: The minimum number of pressure cycles the transducer can endure and still remain within a specified tolerance.

Linearity: The closeness of a calibration curve to a specified straight line. Linearity is expressed as the maximum deviation of any calibration point on a specified straight line during any one calibration cycle.

Liquefied Natural Gas (LNG): Natural gas that has been refrigerated to cryonic temperatures where the gas condenses into a liquid.

Liquefied Petroleum Gas (LPG): Liquefied Petroleum Gas (LPG) is a mixture of low-boiling hydrocarbons that exists in a liquid state at ambient temperatures when under moderate pressures (less than 1.5 MPa or 200 psi). LPG is a by-product from the processing of natural gas and from petroleum refining. Major components of LPG are propane (min. 85% content in the U.S.), butane and propylene.

Liquid Junction Potential: The potential difference existing between a liquid-liquid boundary. The sign and size of this potential depends on the composition of the liquids and the type of junction used.

Load Impedance: The impedance presented to the output terminals of a transducer by the associated external circuitry.

Load: The electrical demand of a process expressed as power (watts), current (amps) or resistance (ohms).

Logarithmic Scale: A method of displaying data (in powers of ten) to yield maximum range while keeping resolution at the low end of the scale.

Loop Resistance: The total resistance of a thermocouple circuit caused by the resistance of the thermocouple wire. Usually used in reference to analog pyrometers which have typical loop resistance requirements of 10 ohms.

Mandrel (Balancing Arbor): An accurately machined shaft on which work is mounted for balancing.

Manual Reset (Adjustment): The adjustment on a proportioning controller which shifts the proportioning band in relationship to the set point to eliminate droop or offset errors.

Manual Reset (Switch): The switch in a limit controller that manually resets the controller after the limit has been exceeded.

Mass Flow Rate: Volumetric flowrate times density, i.e. pounds per hour or kilograms per minute.

Maximum Elongation: The strain value where a deviation of more than ±5% occurs with respect to the mean characteristic (diagram of resistance change vs strain).

Maximum Operating Temperature: The maximum temperature at which an instrument or sensor can be safely operated.

Maximum Power Rating: The maximum power in watts that a device can safely handle.

Mean Temperature: The average of the maximum and minimum temperature of a processequilibrium.

Measurand: A physical quantity, property, or condition which is measured.

Measuring Junction: The thermocouple junction referred to as the hot junction that is used to measure an unknown temperature.

Mechanical Hysteresis: The difference of the indication with increasing and decreasing strain loading, at identical strain values of the specimen.

Medium Effect (f m): For solvents other than water the medium effect is the activity coefficient related to the standard state in water at zero concentration. It reflects differences in the electrostatic and chemical interactions of the ions with the molecules of various solvents. Solvation is the most significant interaction.

Melting Point: The temperature at which a substance transforms from a solid phase to a liquid phase.

Membrane: The pH-sensitive glass bulb is the membrane across which the potential difference due to the formation of double layers with ion-exchange properties on the two swollen glass surfaces is developed. The membrane makes contact with and separates the internal element and filling solution from the sample solution.

Method of Correction: A procedure whereby the mass distribution of a rotor is adjusted to reduce unbalance, or vibration due to unbalance, to an acceptable value. Corrections are usually made by adding material to, or removing it from, the rotor.

Mica: A transparent mineral used as window material in high-temperature ovens.

Microamp: One millionth of an ampere, 10-6 amps, µA.

Micron: One millionth of a meter, 10-6 meters.

Microvolt: One millionth of a volt, 10-6 volts.

Mil: One thousandth of an inch (.001”).

Milliamp: One thousandth of an amp, 10-3 amps, symbol mA.

Millimeter: One thousandth of a meter, symbol mm.

Millivolt: Unit of electromotive force. It is the difference in potential required to make a current of 1 millampere flow through a resistance of 1 ohm; one thousandth of a volt, symbol mV.

Mineral-insulated Thermocouple: A type of thermocouple cable which has an outer metal sheath and mineral (magnesium oxide) insulation inside separating a pair of thermocouple wires from themselves and from the outer sheath. This cable is usually drawn down to compact the mineral insulation and is available in diameters from .375 to .010 inches. It is ideally suited for high-temperature and severe-duty applications.

Mounting Error: The error resultant from installing the transducer, both electrical and mechanical.

Natural Gas (NG): A mixture of hydrocarbon compounds and small quantities of various non hydrocarbon components existing in the gas phase or in solution with crude oil in natural underground reservoirs. The main component of natural gas is methane.

NBS: National Bureau of Standards.

NEC: National Electric Codes.

Negative Temperature Coefficient: A decrease in resistance with an increase in temperature.

Nicrosil/Nisil: A nickel chrome/nickel silicone thermal alloy used to measure high temperatures. Inconsistencies in thermoelectric voltages exist in these alloys with respect to the wire gage.

Nitrogen Oxides (NOx): Several air-polluting gases composed of nitrogen and oxygen which play an important role in the formation of photochemical smog. Nitrogen oxides are collectively referred to as "NOx", where "x" represents a changing proportion of oxygen to nitrogen. Internal combustion engines are significant contributors to the worldwide nitrogen oxide emissions. For the purpose of emission regulations, NOx is composed of colorless nitric oxide (NO), and the reddish-brown, very toxic and reactive nitrogen dioxide (NO2). Other nitrogen oxides, such as nitrous oxide N2O (the anesthetic "laughing gas"), are not regulated emissions.

NMHC: Non-Methane Hydrocarbons.

NMOG: Non-Methane Organic Gases.

Normal (axial) Stress: The force per unit area on a given plane within a body a = F/A

NPT: National Pipe Thread.

Null: A condition, such as balance, which results in a minimum absolute value of output.

O.D.: Outside diameter.

Octal: Pertaining to a base 8 number system.

Offset: The difference in temperature between the set point and the actual process temperature. Also, referred to as droop.

ofhc: Oxygen-free high-conductivity copper. The industrial designation of the pure copper used in a Type T thermocouple.

Ohmeter: An instrument used to measure electrical resistance.

On/off Controller: A controller whose action is fully on or fully off.

On-Board Diagnostics (OBD): A system on board of the vehicle that monitors emission control components and alerts the driver (e.g., by a dashboard light) if malfunctions or emission deterioration occurs. The OBD system involves a number of sensors and a data processor, which is typically integrated with the vehicle's electronic management system.

Open Frontal Area (OFA): In monolith catalyst substrates, the part of the total substrate cross-section area which is available for the flow of gas. The OFA is frequently expressed as a percentage of the total substrate cross-section and sometimes also called the substrate void fraction.

Original Equipment Manufacturer (OEM): Manufacturers of equipment (such as engines, vehicles, etc.) that provide the original product design and materials for its assembly and manufacture. OEMs are directly responsible for manufacturing and modifying the products, making them commercially available, and providing the warranty.

Outboard Rotor: A two-journal rotor which has its center of gravity between the journals.

Overhead Cam: A camshaft used for operating both valves and unit injectors, located on top of or within the cylinder head. Such camshafts are driven by a multi-gear geartrain off the crankshaft. They simplify the design of the cylinder head and eliminate pushrods, allowing for much larger, open intake and exhaust ports and better breathing.

Overshoot: The number of degrees that a process exceeds the set point temperature when coming up to the set point temperature.

Oxygen Storage Capacity (OSC): A capacity of the catalyst washcoat to store oxygen at lean and to release it at rich condition. Typically provided by cerium oxide (ceria), which oscillates between an oxidized and reduced state, depending on the exhaust gas chemistry. The OSC is an important component of three-way catalysts, used to extend the catalyst window.

Oxygenated Fuel: Any fuel substance containing oxygen, such as ethanol, methanol, or biodiesel. Oxygenated fuel tends to give a more complete combustion of its carbon into carbon dioxide (CO2), thereby reducing emissions of hydrocarbons and carbon monoxide. Oxygenated fuels may result in increased nitrogen oxides emissions.

Ozone (O3): An oxygen molecule with three oxygen atoms. The stratosphere ozone layer, which is a concentration of ozone molecules located at 10 to 50 kilometers above sea level, is in a state of dynamic equilibrium. Oxygen molecules absorb ultraviolet (UV) light to form ozone which, in turn, decomposes back to oxygen. These processes absorb most of the ultraviolet light from the sun, shielding life from the harmful effects of UV radiation. Ozone is normally present at ground level in low concentrations. In cities where high level of air pollutants is present, the action of the sun's ultraviolet light can, through a complex series of reactions, produces harmful concentrations of the ground level ozone. The resulting air pollution is known as photochemical smog.

Parallax: An optical illusion which occurs in analog meters and causes reading errors. It occurs when the viewing eye is not in the same plane, perpendicular to the meter face, as the indicating needle.

Perfectly Balanced Rotor: A rotor is perfectly balanced when its mass distribution is such that it transmits no vibratory force or motion to its bearings as a result of centrifugal forces.

Particulate Matter (PM): Particles formed by incomplete combustion of fuel. Compression ignition (diesel) engines generate significantly higher PM emissions than spark ignited engines. The particles are composed of elemental carbon, heavy hydrocarbons (SOF), and hydrated sulfuric acid ("sulfate particulates").

Petroleum: A generic term applied to oil and oil products in all forms. Examples include crude oil, lease condensate, unfinished oil, refined petroleum products, and natural gas plant liquids.

Plane Separation: Of a balancing machine, is the operation of reducing the correction plane interference ratio for a particular rotor.

Poisson Ratio: The ratio between the strain of expansion in the direction of force and the strain of contraction perpendicular to that force v = -Et/E1.

Polarity: In electricity, the quality of having two oppositely charged poles, one positive one negative.

Polycyclic Organic Matter (POM): A class of air toxics defined in the US Clean Air Act as compounds with more than one benzene ring and a boiling point of 100°C and higher. Includes practically all of diesel PAH material.

Polynuclear Aromatic Hydrocarbons (PAH): Aromatic hydrocarbons with two or more (up to five or six) benzene rings joined in various, more or less clustered forms.

Positive Temperature Coefficient: An increase in resistance due to an increase in temperature.

Potential Energy: Energy related to the position or height above a place to which fluid could possibly flow.

Pour Point: A measure of the ability of a diesel fuel to operate under cold weather conditions. Defined as the temperature at which the amount of wax out of solution is sufficient to gel the fuel when tested under standard conditions (ASTM D97).

Precombustion Chamber: A small, auxiliary combustion chamber connected by a narrow orifice with the main chamber. Fuel is injected into the prechamber and ignites there, causing hot gases to expand into the main chamber (cylinder).

Primary Device: Part of a flowmeter which is mounted internally or externally to the fluid conduit and produces a signal corresponding to the flowrate and from which the flow may be determined.

Primary Standard (NBS): The standard reference units and physical constants maintained by the National Bureau of Standards upon which all measurement units in the United States are based.

Principal Axes: The axes of maximum and minimum normal stress.

Probe: A generic term that is used to describe many types of temperature sensors.

Process Meter: A panel meter with sizeable zero and span adjustment capabilities, which can be scaled for readout in engineering units for signals such as 4-20 mA, 10-50 mA and 1-5 V.

Propane (C3H8): A normally gaseous straight-chain hydrocarbon. Propane is a colorless paraffinic gas that boils at a temperature of -42°C. It is extracted from natural gas or refinery gas streams.

Protection Head: An enclosure usually made out of metal at the end of a heater or probe where connections are made.

Protection Tube: A metal or ceramic tube, closed at one end into which a temperature sensor is inserted. The tube protects the sensor from the medium into which it is inserted.

Protocol: A formal definition that describes how data is to be exchanged.

PSIA: Pounds per square inch absolute. Pressure referenced to a vacuum.

PSID: Pounds per square inch differential. Pressure difference between two points.

PSIG: Pound per square inch gage. Pressure referenced to ambient air pressure.

PSIS: Pounds per square inch standard. Pressure referenced to a standard atmosphere.

Pump-Line-Nozzle Fuel System: A fuel system using a single injection pump driven off the geartrain on the front of the engine that also drives the camshaft. The central injection pump feeds a separate injection nozzle located in the cylinder head above each cylinder. Lines which must be of exactly equal length link each pump plunger with the associated nozzle. Each nozzle incorporates a needle valve and the orifices which actually handle atomization.

Purge Test: A test used to determine if fuel vapors are properly drawn from the evaporative canister and the fuel tank into the engine for combustion. If the purge system is not working properly, the evaporative canister can become saturated and vent hydrocarbons into the atmosphere.

Radiation: See Infrared

Rangeability: The ratio of the maximum flowrate to the minimum flowrate of a meter.

Rankine (°R): An absolute temperature scale based upon the Fahrenheit scale with 180° between the ice point and boiling point of water. 459.67°R = 0°F.

Rate Action: The derivative function of a temperature controller.

Rate time: The time interval over which the system temperature is sampled for the derivative function.

Real Time: The time interval over which the system temperature is sampled for the derivative function.

Record: A collection of unrelated information that is treated as a single unit.

Reference Junction: The cold junction in a thermocouple circuit which is held at a stable known temperature. The standard reference temperature is 0°C (32°F). However, other temperatures can be used.

Reference Mark: Any diagnostic point or mark which can be used to relate a position during rotation of a part to its location when stopped.

Reference Plane: Any plane perpendicular to the shaft axis to which an amount of unbalance is referred.

Reformulated Gasoline (RFG): Gasoline whose composition has been changed, from that of gasolines sold in 1990, to (1) include oxygenates, (2) reduce the content of olefins, aromatics and volatile components, and (3) reduce the content of heavy hydrocarbons to meet performance specifications for ozone-forming tendency and for release of toxic substances (benzene, formaldehyde, acetaldehyde, 1,3-butadiene, and polycyclic aromatic hydrocarbons) into the air from both evaporation and tailpipe emissions.

Refractory Metal Thermocouple: A class of thermocouples with melting points above 3600°F. The most common are made from tungsten and tungsten/rhenium alloys Types G and C. They can be used for measuring high temperatures up to 4000°F (2200°C) in non-oxidizing, inert, or vacuum environments.

Register: A storage device with a specific capacity, such as a bit, byte or word.

Relay (Mechanical): An electromechanical device that completes or interrupts a circuit by physically moving electrical contacts into contact with each other.

Relay (Solid State): A solid state switching device which completes or interrupts a circuit electrically with no moving parts.

Renewable Energy: Energy obtained from sources that are essentially inexhaustible, unlike fossil fuels. It includes conventional hydro-electric, wood, biofeedstocks, waste, geothermal, wind, photovoltaic, and solar thermal energy.

Residual (Final) Unbalance: Residual unbalance is that unbalance of any kind that remains after balancing.

Resistance Ratio Characteristic: For thermistors, the ratio of the resistance of the thermistor at 25°C to the resistance at 125°C.

Resistance Temperature Characteristic: A relationship between a thermistor's resistance and the temperature.

Respirable Combustible Dust (RCD): A method of measuring ambient DPM exposures using a combustion process. Used in underground mines in Canada.

Reynolds Number: The ratio of inertial and viscous forces in a fluid defined by the formula Re = rVD/μ, where

r = Density of fluid, μ = Viscosity in centipoise (CP), V = Velocity, and D = Inside diameter of pipe.

Rigid Rotor: A rotor is considered rigid when it can be corrected in any two (arbitrarily selected) planes and after that correction, its unbalance does not significantly exceed the balancing tolerances (relative to the shaft axis) at any speed up to maximum operating speed and when running under conditions which approximate closely to those of the final supporting system.

Rotary Injection Pump: A lower-cost injection pump used with pump-line-nozzle systems. The pump has a central plunger system (usually consisting of two opposing plungers) that provides fuel to every cylinder during the required injection period. A plate located near the top of the pump rotates, opening an appropriate orifice at the right time for distribution to each cylinder's injection nozzle through a separate line. It is usually used with automotive or agricultural engines that have lower performance and durability requirements than the heavy-duty truck diesels.

Rotor: A rotor is a rotating body whose journals are supported by bearings.

RTD: Resistance temperature detector.

SAMA: Scientific Apparatus Makers Association. An association that has issued standards covering platinum, nickel, and copper resistance elements (RTDs).

Selective Catalytic Reduction (SCR): Term frequently used as a synonym for catalytic reduction of NOx in diesel exhaust or flue gases by nitrogen containing compounds, such as ammonia or urea. Such SCR systems are commercially available for stationary applications and are being developed for mobile diesel engines. Since "selective catalytic reduction" is a generic term which also applies to other reactions, its use may lead to confusion in some situations.

Shear Modulus: The ratio of the shear stress and the angular shear distortion.

Shear Stress: Where normal stress is perpendicular to the designated plane, shear stress is parallel to the plane.

Shearing Strain: A measure of angular distortion also directly measurable, but not as easily as axial strain.

Sheath Thermocouple: A thermocouple made out of mineral-insulated thermocouple cable which has an outer metal sheath.

SI: System Internationale. The name given to the standard metric system of units.

Single-Plane (Static) Balancing Machine: A single plane balancing machine is a gravitational or centrifugal balancing machine that provides information for accomplishing single plane balancing.

Size Selective Sampling (SSS): A method of measuring ambient DPM exposures through a mechanical separation (inertial impaction) of particles into two fractions: above and below (typically) 0.8 µm, followed by gravimetric analysis. Based on air quality research in underground mining, it is believed that the particle fraction below the 0.8 µm cut point can be attributed to diesel exhaust.

Smallest Bending Radius: The smallest radius that a strain gage can withstand in one direction, without special treatment, without suffering visible damage.

Soluble Organic Fraction (SOF): The organic fraction of diesel particulates. SOF includes heavy hydrocarbons derived from the fuel and from the engine lubricating oil. The term "soluble" originates from the analytical method used to measure SOF which is based on extraction of particulate matter samples using organic solvents.

Span Adjustment: The ability to adjust the gain of a process or strain meter so that a specified display span in engineering units corr sponds to a specified signal span. For instance, a display span of 200°F may correspond to the 16 mA span of a 4-20 mA transmitter signal.

Span: The difference between the upper and lower limits of a range expressed in the same units as the range.

Specific Gravity: The ratio of mass of any material to the mass of the same volume of pure water at 4°C.

Specific Heat: The ratio of thermal energy required to raise the temperature of a body 1° to the thermal energy required to raise an equal mass of water 1°.

Spot Size: The diameter of the circle formed by the cross section of the field of view of an optical instrument at a given distance.

Spurious Error: Random or erratic malfunction.

SSR: Solid state relay.

Stability: The quality of an instrument or sensor to maintain a consistent output when a constant input is applied.

Stagnation Pressure: The sum of the static and dynamic pressure.

Static Calibration: A calibration recording pressure versus output at fixed points at room temperature.

Static Error Band: The error band applicable at room temperature.

Static Pressure: Pressure of a fluid whether in motion or at rest. It can be sensed in a small hole drilled perpendicular to and flush with the flow boundaries so as not to disturb the fluid in any way.

Static Unbalance: Static unbalance is that condition of unbalance for which the central principal axis is displayed only parallel to the shaft axis

Steady Flow: A flow rate in the measuring section of a flow line that does not vary significantly with time.

Steady State Vibration: That condition of vibration induced by an unchanging continuing periodic force.

Stiffness: The ratio of the force required to create a certain deflection or movement of a part expressed as (Force/deflection) lbs/in or grams/cm.

Strain Gage: A measuring element for converting force, pressure, tension, etc., into an electrical signal.

Strain: The ratio of the change in length to the initial unstressed reference length.

Strouhal Number: A nondimensional parameter important in vortex meter design defined as

s = Fh/V where f = frequency, V = velocity, and h = a reference length

Super Cooling: The cooling of a liquid below its freezing temperature without the formation of the solid phase.

Super Heating: 1. The heating of a liquid above its boiling temperature without the formation of the gaseous phase. 2. The heating of the gaseous phase considerably above the boiling-point temperature to improve the thermodynamic efficiency of a system.

Super Ultra Low Emission Vehicle (SULEV): A vehicle that meets the SULEV emission standards set by the California Air Resources Board. The SULEV standards are more stringent than the ULEV standards.

Supercharger: A compressor device to compress the combustion air or the air/fuel mixture before it enters the engine cylinder. Superchargers are typically driven by the engine itself, through a system of gears, a belt drive, or by an electrical motor.

Swirl Combustion: A combustion chamber configuration which uses curved mixing ridges in the intake ports and/or a re-entrant piston bowl (a bowl whose top edges curve inward). Some swirl combustion chambers have a larger rim around the outside of the piston and a more compact combustion chamber or bowl. The swirl is used to reduce particulate emissions.

Temperature Range, Compensated: The range of ambient temperatures within which all tolerances specified for Thermal Zero Shift and Thermal Sensitivity Shift are applicable (temperature error).

Temperature Range, Operable: The range of ambient temperatures, given by their extremes, within which the transducer may be operated. Exceeding compensated range may require recalibration.

Thermal Conductivity: The property of a material to conduct heat in the form of thermal energy.

Thermal Expansion: An increase in size due to an increase in temperature expressed in units of an increase in length or increase in size per degree, i.e. inches/inch/degree C.

Thermal Gradient: The distribution of a differential temperature through a body or across a surface.

Thermal Sensitivity Shift: The sensitivity shift due to changes of the ambient temperature from room temperature to the specified limits of the compensated temperature range.

Thermal Zero Shift: An error due to changes in ambient temperature in which the zero pressure output shifts. Thus, the entire calibration curve moves in a parallel displacement.

Thermistor: A temperature-sensing element composed of sintered semiconductor material which exhibits a large change in resistance proportional to a small change in temperature. Thermistors usually have negative temperature coefficients.

Thermowell: A closed-end tube designed to protect temperature sensors from harsh environments, high pressure, and flows. They can be installed into a system by pipe thread or welded flange and are usually made of corrosion-resistant metal or ceramic material depending upon the application.

Threshold Limit Value (TLV): Time-weighted average concentration of an air pollutant at the workplace for a conventional 8-hour workday and a 40-hour workweek, to which nearly all workers may be repeatedly exposed without adverse health effects.

Total Carbon (TC): The sum of the elemental carbon and organic carbon associated with diesel particulates. Typically amounts to 80-85% of the total DPM mass.

Total Particulate Matter (TPM): The total particulate matter emissions including all fractions of diesel particulates, i.e. the carbonaceous, organic (SOF), and sulfate particulates.

Transducer Vibration: Generally, any device which converts movement, either shock or steady state vibration, into an electrical signal proportional to the movement; a sensor.

Transducer: A device (or medium) that converts energy from one form to another. The term is generally applied to devices that take physical phenomenon (pressure, temperature, humidity, flow, etc.) and convert it to an electrical signal.

Transient Vibration: A temporary vibration or movement of a mechanical system.

Transitional Flow: Flow between laminar and turbulent flow, usually between a pipe Reynolds number of 2000 and 4000.

Triple Point: The temperature and pressure at which solid, liquid, and gas phases of a given substance are all present simultaneously in varying amounts. The triple point of water is .01°C.

Turbine: A rotary machine which extracts mechanical shaft power from the working fluid (gas or liquid) using rotor vanes.

Turbo Lag: The time delay between injecting fuel to accelerate and delivering air to the intake manifold by the turbocharger. This phenomenon may cause black smoke emissions in some turbocharged diesel engines during accelera-

tion.

Turbocharging: A process of compressing the engine intake air charge in order to allow more air and fuel into the cylinder and, thus, to increase the engine power output. The compressor, called the turbocharger, is driven by an exhaust gas propelled turbine.

Turbo Generator: Gas turbine combined with an electrical generator.

Turbulent Flow: When forces due to inertia are more significant than forces due to viscosity. This typically occurs with a Reynolds number in excess of 4000.

Typical: Error is within plus or minus one standard deviation (±1%) of the nominal specified value, as computed from the total population.

UL: Underwriters Laboratories, Inc. An independent laboratory that establishes standards for commercial and industrial products.

Ultraviolet: That portion of the electromagnetic spectrum below blue light (380 nanometers).

Unbalance: That condition which exists in a rotor when vibratory force or motion is imparted to its bearings as a result of centrifugal forces.

Unbalance Tolerance: The unbalance tolerance with respect to a radial plane (measuring plane or correction plane) is that amount of unbalance which is specified as the maximum below which the state of unbalance is considered acceptable.

Undershoot: The difference in temperature between the temperature a process goes to, below the set point, after the cooling cycle is turned off and the set point temperature.

Union: A form of pipe fitting where two extension pipes are joined at a separable coupling.

Unit Injector: An injector which is camshaft-driven and incorporates a plunger. The plunger works in conjunction with orifices in the injector body to determine the beginning and end of injection. The plungers has a helix and is rotated by a rack so the beginning and end of injection can occur closer together or farther apart, thus shortening or lengthening the injection period and changing metering. A special pump which supplies fuel through an orifice to the injector operates at a pressure which is precisely controlled depending on the changes in engine speed.

Vacuum: Any pressure less than atmospheric pressure.

Velocity: The time rate of change of displacement; dx/dt.

Viscosity: The inherent resistance of a substance to flow.

VMT: Vehicle miles traveled.

Volatile Organic Compounds (VOC): Hydrocarbon-based emissions released through evaporation or combustion. The term VOC is usually used in regard to stationary emission sources.

Volatile Organic Fraction (VOF): The organic fraction of diesel particulate matter as determined by vacuum evaporation. It may or may not be equivalent to the SOF fraction. Depending on the exact analytical procedure, the VOF may include the organic material (SOF) as well as some of the sulfate particulates which, being composed primarily of hydrated sulfuric acid, are also volatile.

Volume Flow Rate: Calculated using the area of the full closed conduit and the average fluid velocity in the form, Q = V x A, to arrive at the total volume quantity of flow. Q = volumetric flowrate, V = average fluid velocity, and A = cross sectional area of the pipe.

Wheatstone Bridge: A network of four resistances, an emf source, and a galvanometer connected such that when the four resistances are matched, the galvanometer will show a zero deflection or "null" reading.

White Smoke: The smoke emitted during a cold start from a diesel engine, composed mainly of unburnt fuel and particulate matter.

Working Standard: A standard of unit measurement calibrated from either a primary or secondary standard which is used to calibrate other devices or make comparison measurements.

Young's Modulus: Young's Modulus (the Modulus of Elasticity) is equivalent to the ratio of normal stress to strain.

Zero Adjustment: The ability to adjust the display of a process or strain meter so that zero on the display corresponds to a non-zero signal, such as 4 mA, 10 mA, or 1 V dc. The adjustment range is normally expressed in counts.

Zero Offset: 1. The difference expressed in degrees between true zero and an indication given by a measuring instrument. 2. See Zero Suppression

Zero Suppression: The span of an indicator or chart recorder may be offset from zero (zero suppressed) such that neither limit of the span will be zero. For example, a temperature recorder which records a 100° span from 400° to 500° is said to have 400° zero suppression.

READING LIST

Armytage, W. H. G. *A social history of engineering.* 2nd ed. London, Faber, 1966.

Billington, David P. *The innovators: the engineering pioneers who made America modern.* New York, Wiley, c1996.

Bruno, Leonard C. *The tradition of technology: landmarks of Western technology in the collections of the Library of Congress.* Washington, Library of Congress, 1995.

Burstall, Aubrey Frederic. *A history of technical engineering.* London, Faber and Faber, 1963.

Channell, David F. *The history of engineering science: an annotated bibliography.* New York, Garland, 1989.

"Company says it has developed cleaner burning diesel fuel".http://www.cnn.com/us/9903/24/cleaner.diesel. [February 1, 2003].

De Camp, L. Sprague. *The ancient engineers.* Cambridge, Mass., MIT Press, 1970, c1963.

"Diesel Engine Exhaust Emissions". http://www.hse.gov.uk/pubns/indg286.htm. [January 31, 2003].

"Diesel Emissions: an exhausting hazard". http://www.whsc.on.ca/Publications/hazardbulletins/fall2001/diesel.htm. [February 8, 2003] "Diesel Exhaust".

http://www1.umn.edu/eoh/hazards/hazards%20site/diesel%20exhaust/diesel-character.html. [February 2, 2003].

"Environmental Health Effects of Diesel Engines" http://www.vpirg.org/campaigns/environmentalHealth/dieselExhaust.html. [February 4, 2003].

Finch, James Kip. *Engineering and Western civilization.* New York, McGraw-Hill, 1951.

Finch, James Kip. *The story of engineering.* Garden City, N.Y., Doubleday, 1960.

Garrison, Ervan G. *A history of engineering and technology: artful methods.* 2nd ed. Boca Raton, Fla., CRC Press, 1999.

Gillespie, LaRoux. "Burrs produced by drilling," Bendix Corporation Unclassified Topical Report, BDX-613-1248.

Hawkes, Nigel. *Amazing achievements: a celebration of human ingenuity.* San Diego, Calif., Thunder Press, c1996.

Hill, Donald Routledge. *A history of engineering in classical and medieval times.* London, New York, Routledge, 1996.

Kérisel, Jean. *Down to earth: foundations past and present: the invisible art of the builder.* Rotterdam, Boston, A.A. Balkema, 1987.

Kim, Jinsoo. "Cost minimization of drilling operation by a drilling burr control chart and bayesian statistics," Ph.D. Dissertation, Department of Mechanical Engineering, University of California at Berkeley: 2000.

Kim, Jinsoom Sangkee Min, David A. Dornfeld. "Optimization and control of drilling burr formation of AISI 304L and AISI 4118 based on DBCC," *International Journal of Machine Tools and Manufacture*, Vol. 41, 2001, pp. 923-936.

Kirby, Richard Shelton, and others. *Engineering in history.* New York, McGraw-Hill, 1956.

Kostic, M. "Sadi Carnot's Ingenious Reasoning of Ideal Heat-Engine Reversible Cycles," Proceedings of the 4th IASME/WSEAS International Conference on ENERGY, ENVIRONMENT, ECOSYSTEMS and SUSTAINABLE DEVELOPMENT (EEESD '08), Algarve, Portugal, June 11-13, 2008. In *New Aspects of Energy, Environment, Ecosystems and Sustainable Development* (Editors: T. Panagopoulos, T. Noronha Vaz , M. D. Carlos Antunes), WSEAS Press: 2008.

Krebs, Michelle. "Diesel Revisited". http://www.popsci.com/popsci/auto/arti-

cle/0,12543,198235-1,00.html. [February 4, 2003].

Kurian, Vinson. "Philippines Set to Switch Over to Coco Diesel". http://www.plannet.com/businessline/2001/06/23/stories/07236761.htm. [February 1, 2003].

Langmead, Donald, and Christine Garnaut. *Encyclopedia of architectural and engineering feats*. Santa Barbara, Calif., ABC-CLIO, c2001.

Marshall, Brian. "How Diesel Engines Work." http://www.howstuffworks.com/diesel.htm/printable. [January 31, 2003].

Neuburger, Albert. *The technical arts and sciences of the ancients*. Translated by Henry L.Brose. New York, Barnes & Noble, 1969.

Parkyn, Neil (ed). *The Seventy wonders of the modern world*. New York, Thames & Hudson, 2002.

Parsons, William Barclay. *Engineers and engineering in the Renaissance*. Cambridge, Mass., M.I.T. Press, 1968.

Pletting and Associates Website: http://www.pletting.com/twist.html.

Rae, John, and Rudy Volti. *The engineer in history*. Rev. ed. New York, Peter Lang, 2001.

Shackelford, James. *Introduction to Material Science for Engineers*, 4th ed, New Jersey: Prentice Hall, 1996.

Stewart, Kevin. "Health Effects of Diesel Exhaust". http://www.californialung.org/downloads/diesel010501/ALA-OEHHA_diesel.pdf. [February 6, 2003].

Tobin, James. *Great projects: the epic story of the building of America: from the taming of the Mississippi to the invention of the Internet*. New York, Free Press, c2001.

Turner, Roland and Steven L. Goulden (eds). *Great engineers and pioneers in technology: From antiquity through the Industrial Revolution*. New York, St. Martin's Press, c1981.

"U.S. EPA launches hearings on reducing sulfur in diesel fuel".http://www.cnn.com/2000/US/06/19/diesel.fuel. [February 3, 2003].

Williams, Archibald. *Engineering feats: great achievements simply described*. London, New York, T. Nelson and Sons, 1925.

Yuki, Hasegawa. "Burr in drilling aluminum and prevention of it." SME Technical Paper, MR75-480.

INDEX